RESOURCING AN AGROECOLOGICAL URBANISM

Foregrounding an innovative and radical perspective on food planning, this book makes the case for an agroecological urbanism in which food is a key component in the reinvention of new and just social arrangements and ecological practices.

Building on state-of-the-art and participatory research on farming, urbanism, food policy and advocacy in the field of food system transformation, this book changes the way food planning has been conceptualised to date and invites the reader to fully embrace the transformative potential of an agroecological perspective. Bringing in dialogue from both the rural and urban, the producer and consumer, this book challenges conventional approaches that see them as separate spheres, whose problems can only be solved by a reconnection. Instead, it argues for moving away from a 'food-in-the-city' approach towards an 'urbanism' perspective, in which the economic and spatial processes that currently drive urbanisation will be unpacked and dissected, and new strategies for changing those processes into more equal and just ones are put forward. Drawing on the nascent field of urban political agroecology, this text brings together: i) theoretical re-conceptualisations of urbanism in relation to food planning and the emergence of new agrarian questions, ii) critical analysis of experimental methodologies and performing arts for public dialogue, reflexivity and food sovereignty research, iii) experiences of resourceful land management, including urban land use and land tenure change, and iv) theoretical and practical exploration of post-capitalist economics that bring consumers and producers together to make the case for an agroecological urbanism.

Aimed at advanced students and academics in agroecology, sustainable food planning, urban geography, urban planning and critical food studies, this book will also be of interest to professionals and activists working with food systems in both the Global North and the Global South.

Chiara Tornaghi (1972) is Associate Professor in Urban Food Sovereignty and Resilience at the Centre for Agroecology, Water and Resilience (CAWR), Coventry University, UK. She has a background in Politics (Lauream, State University of Milan, 2001), Sociology (PhD, University of Milano Bicocca, 2005) and Planning (PgCert, University of Newcastle, UK, 2006). Her research interests include grassroots contestation and reappropriation of public space, politics of urban land, political pedagogies, indigenous cosmologies and knowledge of plants as food and medicine, feminist political ecology and urban agroecology. Since 2016 she is the elected Chair of the AESOP Sustainable Food Planning group. Beside academic life, she is also an allotmenteer, a community food grower, and working towards reskilling herself in medical herbalism.

Michiel Dehaene (1971) is Associate Professor in Urbanism at the department of Architecture and Urban Planning at Ghent University where he leads his own research group and teaches courses in urban analysis and design. He holds a master's degree in engineering-architecture (KULeuven 1994), a Master of Architecture in Urban Design (Harvard University 1996) and a PhD in Architecture and Urbanism (KULeuven 2002). His work focusses on sub-urban renewal, the (planning) history of dispersed urban development, sustainable cities and food planning. His long-term research has been structured around the incorporation of urban theories and theories of urbanization within the fields of planning and design, moving away from normative design theory. This includes systematic work on urban development models and territorial strategies that support the agroecological production of food. With Chiara Tornaghi he leads the JPI SUGI 'Urbanising in Place' project on the development of an Agroecological Urbanism.

Food Literacy
[illegible]
[illegible]

Edited by Sandra Notaro

Routledge Studies in Food, Society and the Environment

Voice and Participation in Global Food Politics
Alana Mann

Plant-Based Diets for Succulence and Sustainability
Edited by Kathleen May Kevany

Sustainable Food System Assessment
Lessons from Global Practice
Edited by Alison Blay-Palmer, Damien Conaré, Ken Meter, Amanda Di Battista and Carla Johnston

Raw Veganism
The Philosophy of the Human Diet
Carlo Alvaro

The Bioeconomy Approach
Constraints and Opportunities for Sustainable Development
Udaya Sekhar Nagothu

Resourcing an Agroecological Urbanism
Political, Transformational and Territorial Dimensions
Edited by Chiara Tornaghi and Michiel Dehaene

Food, the Senses and the City
Edited by Ferne Edwards, Roos Gerritsen and Grit Wesser

For more information about this series, please visit: http://www.routledge.com/books/series/RSFSE/

RESOURCING AN AGROECOLOGICAL URBANISM

Political, Transformational and Territorial Dimensions

Edited by

Chiara Tornaghi and Michiel Dehaene

Routledge
Taylor & Francis Group
LONDON AND NEW YORK

earthscan
from Routledge

First published 2021
by Routledge
2 Park Square, Milton Park, Abingdon, Oxon OX14 4RN

and by Routledge
52 Vanderbilt Avenue, New York, NY 10017

Routledge is an imprint of the Taylor & Francis Group, an informa business

British Library Cataloguing-in-Publication Data
A catalogue record for this book is available from the British Library

Library of Congress Cataloging-in-Publication Data
A catalog record has been requested for this book

ISBN: 978-1-138-35967-3 (hbk)
ISBN: 978-1-138-35968-0 (pbk)
ISBN: 978-0-429-43356-6 (ebk)

Typeset in Bembo
by Deanta Global Publishing Services, Chennai, India

CONTENTS

CONTRIBUTORS

Anna Maria Orrù works with food, biomimicry and artistic research, curating performative research and alternate approaches in ecological design. Based in Sweden and Italy, she holds a PhD in Architecture/Artistic Research and is Senior Lecturer at Konstfack – University of Arts, Crafts and Design. In her spare time, she is a beekeeper and takes care of an olive grove.

Barbara Van Dyck is an agroecology activist from Belgium and an associate professor at the Centre for Agroecology, Water and Resilience at Coventry University. She works on science in society questions with a particular interest in the political ecologies of agriculture and food.

Chiara Tornaghi is Associate Professor in Urban Food Sovereignty and Resilience at the Centre for Agroecology, Water and Resilience, Coventry University, UK. A scholar-activist and food grower, her research revolves around feminist political ecology and urban agroecology. Since 2016, she is elected Chair of the AESOP Sustainable Food Planning group.

C.M. Deh-Tor is an academic joint pen name for critical urban scholars Chiara Tornaghi and Michiel Dehaene.

Dagmar Diesner is a PhD student at the Centre for Agroecology, Water and Resilience, Coventry University, UK, studying the social and agroecological relations of self-governed food systems. As a permaculturalist, she co-founded Montagna Viva (Italy), an organisation working with migrants and promoting agro-biodiversity through food cultivation and cultural events.

Debra Solomon is an artist producing public space food system infrastructure and ecological interventions. Collaborating with local communities actualising food-bearing ecosystems, park-like *food forests* in Amsterdam and The Hague, in 2010 Solomon founded Urbaniahoeve, which in Dutch means 'the city as our farmyard' to develop examples of community-stewarded food forests.

Deirdre (Dee) Woods is a food and farming actionist who advocates for good food for all and a just, equitable food system, challenging the systemic barriers that impact marginalised communities and food producers. Her work meets at the nexus of poverty and hunger, human rights, food sovereignty, community development, policy, research, climate and social justice.

Elise Wach is a researcher and food producer who considers herself a 'critical participant' of the agroecology and food sovereignty movements. In addition to her political and participatory research, she runs a food production and reskilling project in England. She aims to contribute to ecological food systems which provide healthy food on an equitable basis.

Ella von der Haide is a German filmmaker, gardener, eco-queer-feminist activist and spatial planning researcher and teacher. Her documentary film series 'Another world is plantable!' features community gardens and their connection to social movements in Argentina, South Africa, Germany and North America: www.communitygarden.de.

Emilio Travieso (DPhil in International Development, Oxford) is a lecturer at the Université Notre Dame d'Haïti (Port-au-Prince) and a visiting researcher at Campion Hall (Oxford).

Giacomo Lepri, graduated in Cultural Anthropology, is president of the Co.r .ag.gio cooperative, which manages 22 ha of public land in Rome. He is operating in the agricultural sector since 2009 and has experience in food processing and as a chef. He has collaborated as researcher and informant in various publications on urban agriculture. He received the 'Real Food Heroes' award (Navdanya International, 2013).

John Taylor is Assistant Professor of Agroecology in the Department of Plant Sciences and Entomology at the University of Rhode Island. His research focuses on the social and agronomic dimensions of urban agriculture. He has a master's in landscape architecture from the University of Michigan and a PhD in crop sciences from the University of Illinois.

Jose Luis Vivero-Pol works at UN World Food Programme on food crises. He is associate researcher in the Centre for the Philosophy of Law at the University of Louvain. His research interests include food valuations and framings (human rights, public good, commons, commodity) and food systems in transition in Western countries and fragile states.

Luca Colombo is the Secretary General of the Italian Foundation for Research in Organic and Biodynamic Agriculture (FIRAB). He has published grey and scientific articles, further to books on agro-biotech and food security issues. His research interests include organic and biodynamic farming, food security and sovereignty, co-innovation processes and methodologies.

Mama D. Ujuaje is an active community researcher, food grower, plant-whisperer, public speaker, transformative workshop facilitator and social educator, engaging many narratives of #foodjustice, #foodknowledges, the body, gender, race, the ways our cultures interact and also fail to, our social aspirations and overlaps and the spaces in-between. www.communitycentredknowledge.org; @indigenousknow

Marjolein Visser currently leads research and teaches fulltime in farming systems and agroecology at the ULB – University of Brussels. Her PhD was on native seed multiplication for the ecological restoration of degraded dry lands, which intertwined agronomical, ecological and social questions.

Mark Tilzey is Associate Professor in the Governance of Food Systems for Resilience, Centre for Agroecology, Water and Resilience, Coventry University, UK. His research interests lie in political ecology, food regimes, agrarian change, agroecology and the international political economy of agri-food systems.

Michiel Dehaene is Associate Professor in Urbanism at the Department of Architecture and Planning, Ghent University, Belgium, where he teaches courses in urban analysis and design. His research focuses on suburban renewal, the (planning) history of dispersed urban development, sustainable cities and food planning.

Nathalie Pipart studied political science and environmental sciences and management at the University of Brussels. She has worked as a researcher in several universities in Belgium in the past years, exploring transdisciplinary approaches for sustainability, with public administrations, urban farmers and forest governance practitioners.

Noémie Maughan graduated in BioEngineering in 2008. Through foreign work experience in development projects (Africa and Latin America), she worked within pluri-disciplinary teams in rural contexts, in close involvement with the field. Being part of the Agroecology Lab team (Belgium) since 2014, she specialises in co-innovation processes in (peri-)urban agriculture.

Priscilla Claeys is Associate Professor in Food Sovereignty, Human Rights and Resilience at the Centre for Agroecology, Water and Resilience (CAWR),

Coventry University (UK). She holds a PhD in Political and Social Sciences from the University of Louvain (2013). Priscilla's main research focus concerns human rights and social movements.

Robin Grey is a musician and social historian who created the show 'Three Acres and a Cow – A History of Land Rights and Protest in Folk Song and Story'. He travels the country teaching people how housing, farming, recreation, climate change, community and nature are affected by Britain's historic land injustices.

Salvatore Engel-Di Mauro is Associate Professor at the Geography Department of SUNY New Paltz (USA). He specialises in issues of soil degradation, urban food production, trace element contamination and eco-socialism. He is chief editor for the journal *Capitalism Nature Socialism*.

Sarah Lovell serves as the H.E. Garrett Endowed Chair Professor and Director of the Center for Agroforestry at University of Missouri, USA. With a focus on the analysis and design of multifunctional landscapes, Dr. Lovell's research programme has emphasised whole-farm planning, productive agroforestry and urban agriculture.

Santiago Ripoll is a social anthropologist specialising in ethnographic and participatory approaches to health and food system analysis, with an emphasis on ethics. He uses an anthropological lens within trans- and inter-disciplinary research in humanitarian emergencies and policy debates around food.

Severin Halder (PhD) is inspired by the everyday resistance of urban peasants in Rio de Janeiro, Bogotá and Maputo. His work is generated from within the popular education collective *orangotango and the Allmende-Kontor Network in Berlin*. His activist geography aims towards the creation of solidarity relationships, horizontal knowledge exchange and self-organised struggles.

Stefano Grando holds an MSc in European Regional Development (Cardiff University) and a PhD in Agrarian Economics (University of Basilicata). He has worked in several EU-funded projects and in monitoring and evaluation of regional development programmes. Currently he works for the Italian Ministry of Agricultural, Food, Forestry and Tourism Policies and for the University of Pisa.

Tomaso Ferrando is Research Professor at the Faculty of Law (Law and Development Research Group) and Center of Development Policy at the University of Antwerp (Belgium). His academic work focuses on the legal destruction and construction of ecological and regenerative food systems. Through the Global Legal Action Network, he provides legal support to communities and peoples affected by land grabbing.

PREFACE FROM THE CHAIR OF AESOP SUSTAINABLE FOOD PLANNING GROUP

Chiara Tornaghi

This book gathers together a selection of contributions presented at the 8th Annual Conference of the AESOP Sustainable Food Planning Group, hosted by the Centre for Agroecology, Water and Resilience (Coventry University), in Coventry, UK, in November 2017. Titled 'Re-imagining sustainable food planning, building resourcefulness: food movements, insurgent planning and heterodox economics', the conference aimed to steer the attention of the AESOP Sustainable Food Planning community towards the radically transformative message of political agroecology and its value system.

Two decades from its early conceptualisation, and ten years on from its institutionalisation in the planning community (Van der Valk and Viljoen, 2014), sustainable food planning had become a thriving transdisciplinary research and policy field bringing together policy-makers, academics and practitioners across the globe. Food charters, food strategies and food policy councils have multiplied, 'alternative food networks' have gained significant and growing shares of the food market and new forms of localisation of food production, including urban agriculture, have been gaining ground and have become central components of new food policy strategies.

However, the scale and speed of the 'food' crisis rendered these achievements modest and utterly inadequate. Urban food poverty and malnutrition, and the related use of food banks, have been on the rise, even in some of the wealthiest countries of the world; the most vulnerable populations in both the Global North and the Global South are left unshielded by austerity politics, food-commodity speculation, land grabbing, or staple food price rises. Diet-related diseases (such as diabetes and obesity) have been growing at alarming rates among children in the supposedly 'well-fed' countries of the world. The world still wastes between 30% and 50% of the food produced, while millions of farmers and land workers growing food across the globe are struggling to make a living. The environmental impacts of the food 'regime' and its diets are plainly devastating.

Planning for sustainable food production and food provision is more than ever urging us to look for more effective, equitable and just approaches that radically change not only the way we grow food, but also the very core of our living space.

The aim of the conference, then, was to discuss ideas, approaches and practices that could help to reinvent food planning in light of the aim to build *a resourceful, agroecological urbanism*.

Inspired by a seminal paper from MacKinnon and Derickson (2013), we used the term 'resourceful' as a particular way of intending the concept of 'resilience': an urbanism that creates the conditions for its inhabitants to control the means of their social reproduction, to have a say on, or direct control of, the resources for their own survival; a space where land, water and nutrients serve the needs of the people (rather than profit) while respecting the ecosystem. A 'resourceful' urbanism creates living conditions that enable people to be resilient while at the same time challenging the root causes of the crisis that require us to look for resilience.

By 'agroecological' we explicitly referred to practices aligned to 'peasant agroecology' and the agroecology movement: a way of cultivating the soil, managing ecological relationships and disposing of the produce that respects the environment and is based on cultural and social arrangements inspired by solidarity and mutuality.

By 'urbanism' we referred to more than just buildings, zoning or planning, and more than just urban contexts. We referred to ensembles of the built environment and its regulation, the material infrastructure and the collective arrangements (for food provision, waste collection, land management, urban design, housing, energy and so forth) that are in place and reflect a specific ideological, social and economic arrangement that is historically specific. In the context of planetary urbanisation and neoliberal capitalism systems, in which rural and urban realms are ruled by the same socio-economic and ideological visions, we aimed to reflect on the urban, the peri-urban and the rural realm simultaneously, and to reflect on their mutual interconnections and dependencies.

While food entered the planning agenda more than a decade ago, a resourceful and agroecological *urbanism*, which is more than closing metabolic loops through urban agriculture, is yet to be conceptualised. An urbanism in which food is not the latest 'fix' to be added as a new way to market but rather a key and long forgotten component around which new and just social arrangements, ecological practices and ways of life must be reinvented.

The selected contributions in this book valorise and bring to the fore a multiplicity of heterodox experiences, policies, concepts and practices that are creating new worlds in innovative and socially just ways, and bear the potential of becoming building blocks of sustainable food planning for a resourceful, agro-ecological urbanism. They also include critical reflections on how current main-stream approaches to food production, food strategies and urban agriculture can be/are being radically transformed into tools for resourcefulness.

I believe the book will be a valuable source for the Sustainable Food Planning community, to reflect on its path, to embrace the challenge of radical food system transformation, and to contribute to fostering and building socially and ecologically just worlds.

Coventry, July 2020

References

MacKinnon, D. and Derickson, K. D. (2013) From Resilience to Resourcefulness: A Critique of Resilience Policy and Activism, *Progress in Human Geography*, 37(2), pp. 251–268.

Van der Valk, A. and Viljoen, A. (2014) AESOP's Thematic Groups - Part 3: The Sustainable Food Planning Thematic Group, *disP - The Planning Review*, 50(4), pp. 78–82, doi: 10.1080/02513625.2014.1007647.

INTRODUCTION

Embracing political agroecology, transforming sustainable food planning

Chiara Tornaghi and Michiel Dehaene

This book foregrounds innovative and radical approaches on sustainable food planning, built from the perspective of agroecology, and makes the case for an 'agroecological urbanism'.

Building on state-of-the-art and participatory research on farming, urbanism, food policy and advocacy in the field of food system transformation, this book aims to change the way food planning has been conceptualised to date, and to enable the reader to fully embrace the transformative potential of an agroecological perspective.

With the progressive co-optation of debates and practices dedicated to rethinking the link between planning and food systems to the services of a new wave of capitalism restructuring and expansion (see, for example, the realms of 'sustainable' development, nature-based 'solutions', green 'growth', green gentrification and the new wave of high-tech and soil-less urban agriculture, etc.), we feel that a book that makes a radical stance and reorients the debate is needed.

In keeping in dialogue the rural/urban and the producer/consumers realms, the book also challenges conventional approaches that see them as separate spheres whose problems can be solved by a reconnection. The book rather makes the case for moving away from a 'food-in-the-city' approach, typical of food planning as a discipline, towards an 'urbanism' perspective, in which the economic and spatial processes that currently drive urbanisation will be unpacked and dissected, and new strategies for changing those economic and spatial processes into more equal and just ones put forward. In doing so, and therefore in moving from food planning to exploring new models of urbanisation and livelihoods with social reproduction and ecological considerations at their core, the book makes the case for a resourceful, agroecological urbanism.

1. How political agroecology intersects and challenges sustainable food planning

Encountering political agroecology has been a transformative experience that has inspired and steered the journey that led to this book. For a number of years, both of us (editors) have been interested in the emergence of food growing and food planning in urban contexts. We embarked upon this journey with two distinct, yet partially overlapping interests. We were looking to explore how urban food production (and the urban handling of nutrients) could become a tool to challenge and reverse planning processes that have dispossessed people from the possibility of controlling the processes and resources that shaped fundamental aspects of their lives, such as the quality of shared urban space and food allocation. We looked into how bottom-up knowledge production and knowledge sharing in relation to food growing, and the direct reclaiming of resources (land, nutrients, water) that were emerging in relation to urban agriculture, could change the politics of urban and peri-urban land (Tornaghi, 2012a, 2012b, 2014a, 2014b, 2017; Tornaghi and Certomà 2018).

We also looked into how the expanding horizontal metropolis, which more and more looks like the destiny of ever-expanding processes of planetary urbanisation, could be amended and rendered less extractive and parasitic towards the environment, and more resourceful for its inhabitants (Dehaene, 2012, 2015). We felt the need to challenge some of the common wisdom of urban planners regarding sustainable urbanism as manifest in the compact city debate (Neuman, 2005), the urban age discourse (Burdett and Sudjic, 2007) and the persistent 'methodological cityism' within urban theory (Angelo and Wachsmuth, 2014). We looked at the emerging academic interest for urban food production and food planning more in general as a promising arena where to find insights and a fertile ground for these endeavours.

Some of the works that have sparked our imagination include the early publications that shaped the debate on sustainable food planning (such as Viljoen, 2005; Sonnino, 2009; Viljoen and Wiskerke, 2012; Steel, 2008), or that highlighted the new solidarities that were emerging through urban gardening (Lyson, 2004; Nordhal, 2009), the critical work that looked at the potential of urban agriculture as a form of agency vis-à-vis debate on autonomy (Hodgkinson, 2005; Saed, 2012) and food injustice (Alkon and Agyeman, 2011; Heynen et al., 2006, 2012; Dixon, 2014), and the debates on urban metabolism and nutrients sovereignty (Castán-Broto et al., 2012; McClintock, 2010; Schneider and McMichael, 2010).

Alongside our academic engagement with these debates, we were also practically involved in, respectively, urban food growing and urban planning, in our different capacities as micro-farmer/community food grower, activist, educator and consultant. These experiences contributed a critical understanding of the deep contradictions that these literatures were only partially able to address.

For example, despite the growing proliferation of urban food growing and the great satisfaction that practitioners felt from this engagement (i.e., in feeling more connected to the local community, caring for their neighbourhood or being more physically active), it was evident that from a food perspective, such as the ability to produce, share and consume food, these were more often than not highly frustrating experiences, constrained by what has been defined as the food-disabling city (Tornaghi, 2017). Urban agriculture was more distinctively a community-building experience, a health promotion strategy, or a form of leisurely reconnection with nature, rather than a way to produce substantial amounts of ecologically sustainable, and socially just, food capable of breaking the dependency from the 'food regime' (Friedmann and McMichael, 1989). It also became clear that despite a growing number of food policy councils, some interesting evolutions on the front of municipal food strategies and food procurement, and the emergence of a food-concerned planning system, processes of urbanisation and collective ways to handle food were still subjected to the same extractive, disempowering and money-driven processes that have shaped urbanisation since the rise of capitalism: think of land speculation and the destruction of use values, and the ongoing loss of small farmers. The new 'urban food' did not substantially change the mechanism for food production (still largely using ordinary artificial fertilisers and pesticides), food commercialisation (still offering farmers extortionately low prices), food allocation (with the most vulnerable populations in full employment yet still in food poverty), or the liberal approach to the (mis)management of urban land and nutrients (with individuals still allowed to use pesticides, and soil nutrients being taken away as waste). The profit-driven ways in which food is extracted from farmers, allocated according to wealth or disposed of as waste, remained largely mainstream. The discovering of food as a new topic for urban planning, and the surging of urban agriculture as a new economic realm, did not seem to lead, unequivocally and systematically, to more equitable forms of land management and just food systems. Innovations seemed to work around constituted interests, and residual opportunities opened up by deindustrialisation and the largely shared approach to food as a commodity. Despite interesting emerging forms of political gardening that were challenging the agency of city-making (Certomà and Tornaghi, 2015), the neoliberal urbanism we knew was distinctively a food-disabling one.

While urban agriculture remained rich in its potential to become an empowering field of practice (Alkon and Agyeman, 2011; Heynen et al., 2012), we also saw dangers in its ongoing co-optation. This was in the form of green gentrification, new market expansions under the remit of circular economies, the attempts to monetise ecosystem services, nature-based solutions and purely technical/industrial translations of the issues at stake (vertical farming, aquaponics, etc.) functional to ongoing speculative approaches to urban land.

While a number of critical scholars started to point out these dangers (Atkinson, 2013; McClintock, 2014; Tornaghi, 2014a), as our sense of frustration within the 'urban agriculture' debate grew, our joint journey started with the aim of searching for a food-empowering urbanism (Deh-Tor, 2017). We did not want to give up the chance to imagine alternative urban solidarities. We felt the need to keep imagining what a radically different urbanism would look like, remaining attached to a shared commitment to the urban as a potentially emancipatory condition. Not only is the urban the condition under which, in all its varieties, half the world's population is living; we believe that the history of urbanism is not simply exhausted by the logics of capitalist oppression that have shaped its dominant face. In the same way that the diverse economies literature (Gibson-Graham, 1996) has tried to break the spell of a totalising critique of capitalism, we feel the need to break the spell of a totalising critique of urbanism, opening our eyes to the existing diversity that contains the germ of radical alternatives.

The discovery of *political agroecology* gave us important tools to begin to question the sustainable food planning that we knew.

The term 'political agroecology' includes a group of practices and approaches, reclaimed and reproduced by a wide group of farmers, scholars and activists across the world. These include agricultural practices respectful of soil ecology and soil health, promoting biodiversity, recognising multi-species solidarities, cherishing horizontal knowledge reproduction and valuing people's knowledge and place/culturally sensitive practices (de Molina, 2012; Van Dyck et al., 2017; Deh-Tor, 2017). Examples of these include companion planting, care for living soils and the regenerative treatment of soil organic matter and nutrient levels, biological management of pests, skills-sharing and solidarity economies.

The attribute 'political' relates to the coalitions, networks, statements and movement-building through which these practices reflect on their condition and reclaim their legitimacy vis-à-vis the aggressive neoliberal and profit-oriented approaches that have attacked, disabled and delegitimised these knowledges, and dispossessed, displaced and marginalised the people who practice them, and who find their livelihood residualised. Political agroecology strongly aligns to the food sovereignty stance against the food regime (the international trade agreements, the oligarchy controlling seeds and agrochemical inputs and the range of policies that favour large agro-industry players over smallholders) that shapes the food system.

It calls for the urgent need to address the ecological crisis of nutrient loss, soil depletion and climate change *at the same time* as addressing the astonishing growth of food and water insecure global (and particularly urban and peri-urban) populations: issues that most food policies address very selectively and in ways largely compliant with the capitalist interests and economies at the root of the problem. This book embraces this challenge.

The specificity of this book is not limited to adopting a political agroecology perspective when thinking sustainable food planning: it goes further and it unfolds by bringing it into the urban realm, and in doing so, contributes to shaping an *urban* political agroecology (cit: van Dyck et al., 2018; Tornaghi and Halder, forthcoming; Tornaghi and Dehaene, 2020).

Given the systemic and global focus of political agroecology, the international nature of the politics and institutions that shape the food regime, the historical metabolic and epistemological rift generated and consolidated through processes of urbanisation, has also meant that translating its heuristic value for considerations on sustainable food systems has brought to light the necessity to move beyond a food-in-the-city approach or a food policy lens. It focuses on questions of urbanism and urbanisation more generally and how they shape our relationship to food.

In order to clarify the focus on urbanism, it is necessary to provide a definition. In the context of the multiplication of 'new urbanisms', the term has increasingly been used in a variety of ways, swinging from meanings close to its French roots (the term 'urbanisme' and its reference to the disciplines of urban planning and design) to meanings closer to classical sociological traditions as exemplified by the classical text of Louis Wirth 'Urbanism as a way of life' (1938). In this book we use the term 'urbanism' as it emerged in critical Marxist studies in the 1970s (Lefebvre, 1974; Harvey, 1973), where it bears a sum of the two: urbanism is the (always contested) ensemble of social arrangements, lifestyles, divisions of labour, cultural practices and social solidarities that materialise and shape the urban environment through processes of urbanisation. By stressing the link between urbanism and urbanisation, we wish to stress the fact that the urban context is more than just a physical space, but always the manifestation of socionatural, sociocultural and socioeconomic processes and ideas that the disciplines of urban planning and design 'serve'.

Our first message to the community of scholars and practitioners interested in food system transformation, and the departing point of this book, is that radical transformations of the food system that seriously aim to tackle the ecological and social justice challenges at once have to consider the way in which urbanism and urbanisation are entangled with the reproduction of capitalist structures and value systems. Current urbanism tends to serve the *status quo*, alternative urbanism may contribute to food systems' transformation.

A political agroecology informed look at the dynamics of urbanisation therefore intersects and challenges sustainable food planning in a number of ways. For example, it looks at land not as a neutral canvas on which various land use patterns can be interchangeably inscribed but views urban soils as valuable common goods. It looks at the geography of food systems as metabolically interrelated socio-ecological patterns that act across the rural-urban divide. It seeks to understand the multiple relationships of interdependence and solidarity between humans and non-humans, living 'critters' and inanimate things. It looks at the livelihood and the collective valuing of smallholders and seeks to break the anthropocentric and charity charge of food.

2. Political agroecology approaches resourcing an agroecological urbanism

The contributions in this book can be seen as examples of research themes in the emerging field of political agroecology. They have been selected for their strength in thinking about transformation and political trajectories, and in this sense resourcing what we call an 'agroecological urbanism' (cf. Chapter 1).

Many of the chapters address more than one of the following themes simultaneously:

- theoretical re-conceptualisations of urbanism, food and food systems' transformation;
- pedagogies of transformative learning and resistance;
- advocacy perspectives for merging urban and rural food justice movements;
- experiences of resourceful land management, including urban land use and land tenure change;
- landscape agroecology and urban agroforestry;
- experimental policies and practices for land access and urban soil care;
- theoretical and practical exploration of post-capitalist economics;
- heterodox economies of resistance.

We see these diverse reflections and experiences as ways of *resourcing* an agroecological urbanism from a multitude of communities and perspectives. By 'resourcing' we mean providing tools and ideas that can help imagine new transformative politics, building political subjectivities, or strengthening activism around underestimated but crucial policy areas (i.e., soil protection). The diversity of approaches is necessary precisely because urbanisms entails the coordination of interdependency in different spheres of life, and moving beyond a capitalist urbanism requires the building of solidarity and coordination across intersectional, transdisciplinary and decolonial activist practices and knowledge.

While we discuss in more detail how the different contributions in this book contribute to an agroecological urbanism, we offer below a synopsis for each chapter.

Synopsis of the book

In this introduction we have so far given an overview of how political agroecology intersects and challenges sustainable food planning as we know it to date, which questions an agroecological perspective raises, and what innovative dimensions it entails when thinking about food system transformation. In this concluding section, we now offer a detailed synopsis of each chapter in order to explain the rationale for the book and highlight the intellectual trajectory that underpins the quest for an agroecological urbanism.

In Chapter 1, C.M. Deh-Tor make the case for an agroecological urbanism, a concept through which they seek to detach the way urbanisation is structured from the extractive, resource-destroying, food-disabling logics of capitalism and neoliberal urbanism, and try to imagine an urbanism rooted in the core values of political agroecology: solidarity, mutual learning, multi-species (more than human) exchanges, environmental stewardship and people's resourcefulness. Starting from a critical review of the urban agenda of sustainable food planning and its limits, and building on feminist social reproduction theory and its decolonial, non-extractive and post-patriarchal perspective, the authors discuss how urban political agroecology presents food planning with a food-inclusive urban geography, an expanded urban food agenda and a heterodox self-description of the field formerly called planning, that enables the reimagination of an alternative urbanism.

Bringing forward transformative ideas means fostering transformative learning, research and communication. In Chapter 2, Ella von der Haide, Anna Maria Orrù, Barbara Van Dyck, Debra Solomon, Mama D. Ujuaje, Deirdre (Dee) Woods, Severin Halder and Robin Grey give us the opportunity to reflect on how ideas of agroecology and food sovereignty are being debated and researched with the aid of experimental practices. Their contributions offer a starting point to reflect on scholar-activist positionalities and the potential and limitations of new emerging experimental methodologies using performative techniques to promote reflexivity and public dialogue in food sovereignty research.

Rather than being static and politically neutral, food systems are socio-ecological networks that are in continuous transformation and where interactions are defined by the activities of people and the planet as much as by the ideas that legitimise certain behaviours. In Chapter 3, five research activists, Tomaso Ferrando, Priscilla Claeys, Dagmar Diesner, Jose Luis Vivero-Pol and Deirdre (Dee) Woods, join together to discuss concrete examples that show that the agroecological transition could be strengthened by the adoption of a political understanding of commons and commoning as intersectional antidotes for a just agroecological transition, one that rejects the colonial, patriarchal, unjust and anti-ecological premises of the mainstream food systems. Through a combination of theory and practice, history and imagination, empowerment and de-commodification, the chapter brings to the forefront those dimensions of food that cannot be monetised and valued in market terms, showing that political, imaginative and organisational power of commons and commoning can bridge the urban-rural divide and contribute to the convergence of various movements, including agroecological urbanism and food sovereignty.

In Chapter 4, Luca Colombo, Stefano Grando and Giacomo Lepri reflect on the politics of peri-urban land reclamation and redistribution as a tool for resourcefulness. Peri-urban agriculture is receiving growing attention in Italy, as elsewhere in Europe. The quest for land is not just an individual endeavour. New farming initiatives proliferate, combining economic returns with socio-environmental ambitions, giving rise to an increasing demand for land by both farmers and 'neo-rural' people. The chapter reviews a mobilisation exercise advocating access to land in Rome, including the 2013 response of the city and regional administration to make public farmland available. The case study showed that food sovereignty and agroecological farming were seen by both activists and local administrations as a way to gain citizens' consensus and represented criteria for land lease tenders.

Drawing on a background in landscape planning, landscape ecology and regional food systems, in Chapter 5, Sarah Lovell and John Taylor review the ongoing discussion on the relevance of agroforestry within an urban context. Urban agroforestry offers a transformative solution that supports production functions through the provision of healthy fruit and nuts but within a perennial system that closely mimics a multi-strata forest ecosystem. When approached from an agroecological perspective, urban agroforestry could contribute to resilience planning initiatives by improving food security, climate change adaptation and microclimate conditions. Chicago, IL, USA, serves as a case study for considering urban agroforestry applications through both retrofitting existing green spaces and planning future multifunctional landscapes.

In Chapter 6, Salvatore Engel-Di Mauro offers an overview of some of the major challenges urban soils can present to food producers in industrialised cities. Following an overview on soil formation, on agroecological approaches to soil and on the specific nature of urban soils, the author discusses the major ecological challenges of urban food production with a particular emphasis on trace element contamination. Technical measures to deal with pollutants are described and shown to be insufficient without a grasp of the political ramifications of contamination. It is argued that this latter aspect is what agroecological urbanism can help confront more effectively, provided the perspective is firmly grounded in both social and biophysical analytical frameworks and just as firmly committed to overcoming capitalist relationships for an ecologically constructive egalitarian alternative.

In Chapter 7, Noémie Maughan, Nathalie Pipart, Barbara Van Dyck and Marjolein Visser look at neo-farmers in peri-urban market gardens and how they deal with the challenges of putting agroecological ideas into practice. The chapter offers insights into contradictory pressures and forces that need to be overcome to make radical ideas work. The chapter presents the results of a reflexive process aiming to critically disentangle the dynamics behind what the farmers construct as an 'ideal urban agriculture activity', and the compromises they have to make while putting the ideal into practice. The study was carried out with urban farmers as co-researchers in a broader participatory action research project in Brussels. Their case study reveals a paradox between the regional institutions' high expectations of urban agriculture's contributions to a sustainable city, and a financial support system, still shaped in a 'pro-growth' mould, pressing neo-farmers towards short-term economic independence at the cost of a long-term agroecological perspective.

In Chapter 8, Elise Wach and Santiago Ripoll take a closer look at the role of rural farmers (in both the Global North and South) in food system transformation. Their work starts from acknowledging the overwhelming evidence that our food systems are not currently working to nourish our populations, ecosystems, economies, or social connections. Agroecological approaches have been shown as having the potential to address many of these problems in the mainstream food system, particularly when combined with concepts of food sovereignty. The approaches localise control and place producers and consumers at the centre of decision-making. Yet there have been questions about whether producers of varying positionalities would develop strategies that are sufficiently radical to transform existing food systems. The chapter reflects the outcomes of a participatory farmer-led research initiative in which producers from the Global South (Nicaragua) and the Global North (England) developed strategies for transforming their food systems. It shows that farmers' framings of food system problems and proposed solutions became more radical over the course of the participatory process. This confirms that participatory processes with farmers have the potential to transform food systems to become more socially and ecologically regenerative.

In Chapter 9, Emilio Travieso brings us to Chiapas, Mexico, to delve into the strategy of one specific group of farmers across the urban-rural divide. After regaining their ancestral lands in Chiapas from plantation owners in the 1990s, the Mayan Tseltal people, accompanied by Jesuits and other collaborators in the Misión de Bachajón, have continued to build their full sovereignty. The chapter explains how the Misión de Bachajón withdraws certain parts of its economy, such as land, food and natural resources, from market commodification while engaging the market through the coffee value chain. Agroecology articulates the two realms of the economy – market and solidarity economy – and holds them together. The economic system of the Misión de Bachajón contributes to the wider common good, both through its rural-urban linkages and through its conservation of biodiversity. However, it must seek out strategic alliances in order to survive in the midst of an adverse system.

In Chapter 10, Mark Tilzey discusses how urban agroecology, as an expression of food sovereignty, constitutes an important 'beachhead' against capitalism by addressing precarity through counter-hegemonic modes of local food provisioning. Tilzey, however, also makes clear that if this 'beachhead' is to be secured and widened, the 'right to the city' needs to go hand in hand with calls for the 'right to the countryside' as part of a broader counter-hegemonic movement that seeks to 'visibilise' the 'imperial mode of living' and to institute ecological sustainability and food equity as key components. In the Global North, this will not be easy, as it requires nothing less than breaking away from 'resource imperialism'. The Southern precariat, Tilzey argues, has a pivotal role in fomenting transitions to post-capitalism, not merely in the South but by 'closing down' resource imperialism, and thereby disabling the capacity of the Northern state-capital nexus to co-opt counter-hegemony through the 'imperial mode of living'.

The book *concludes* with a section where the editors project the book forward by reinstating the key pillars upon which the programmatic dimension of an agroecological urbanism is built, and by reflecting on the progress that this agenda has made since they started the journey, both in research and in action.

References

Alkon, A. H. and Agyeman, J. (eds) (2011) *Cultivating Food Justice: Race, Class, and Sustainability*. Cambridge: MIT Press.

Angelo, H. and Wachsmuth, D. (2014) Urbanizing urban political ecology: A critique of methodological cityism. *International Journal of Urban and Regional Research*, vol 39, no 1, pp16–27.

Atkinson, A. (2013) Readjusting to reality: Urban and peri-urban agriculture to ease the downward passage. *City: Analysis of Urban Trends, Culture, Theory, Policy, Action*, vol 17, no 1, pp85–96.

Burdett, R. and Sudjic, D. (2007) *The Endless City: The Urban Age Project by the London School of Economics and Deutsche Bank's Alfred Herrhausen Society*. London: Phaidon.

Castán-Broto, V., Allen, A. and Rapoport, E. (2012) Interdisciplinary perspectives on urban metabolism. *Journal of Industrial Ecology*, vol 16, no 6, pp851–861.

Certomà, C. and Tornaghi, C. (2015) Editorial. Political gardening. Transforming cities and political agency. *Local Environment: The international journal of justice and sustainability*, vol 20, no 10, pp1123–1131.

De Molina, M. G. (2012) Agroecology and politics. How to get sustainability? About the necessity for a political agroecology. *Agroecology and Sustainable Food Systems*, vol 37, no 1, pp45–59.

Dehaene, M. (2012) *Gardening the Urban Field*. Ghent: A&S/Books.

Dehaene, M. (2015) From the nebular city to the horizontal Metropolis. In: Uyttenhove, P., Dehaene, M., Beeckmans, L., Michelena, A., Notteboom, B., Peleman, D. and Pelger, D. (eds) *A Landscape Perspective on Urbanism*. Ghent: Academia Press, pp71–90.

Deh-Tor, C. M. (2017) From agriculture in the city to an agroecological urbanism: The transformative pathway of urban (political) agroecology. *Urban Agriculture Magazine*, no 33, pp8–10.

Dixon, B. A. (2014) Learning to see food justice. *Agriculture and Human Values*, vol 31, no 2, pp175–184.

Friedmann, H. and McMichael, P. (1989) Agriculture and the state system: The rise and fall of national agricultures, 1870 to the present. *Sociologia Ruralis*, vol 29, no 2, pp93–117.

Gibson-Graham, J. K. (1996) *The End of Capitalism (As We Knew It): A Feminist Critique of Political Economy*. Oxford: Blackwell.

Harvey, D. (1973) *Social Justice and the City*. Baltimore, MD: Johns Hopkins University Press.

Heynen, N., Kaika, M. and Swyngedouw, E. (eds) (2006) *In the Nature of Cities: Urban Political Ecology and the Politics of Urban Metabolism*. New York: Routledge.

Heynen, N., Kurtz, H. E. and Trauger, A. (2012) Food justice, hunger, and the city. *Geography Compass*, vol 6, no 5, pp304–311.

Hodgkinson, T. (2005) Digging for anarchy. In: Richardson, T. and Kingsbury, N. (eds), *Vista: The Culture and Politics of Gardens*. London: Frances Lincoln, pp66–73.

Lefebvre, H. (1974) *La production de l'espace*. Paris: Anthropos.

Lyson, T. A. (2004) *Civic Agriculture: Reconnecting Farm, Food and Community*. Medford, MA: Tufts University Press.

McClintock, N. (2010) Why farm the city? Theorizing urban agriculture through a lens of metabolic rift. *Cambridge Journal of Regions Economy and Society*, vol 3, no 2, pp191–207.

McClintock, N. (2014) Radical, reformist, and garden-variety neo-liberal: Coming to terms with urban agriculture's contradictions. *Local Environment. The International Journal of Justice and Sustainability*, vol 19, no 2, pp147–171.

Neuman, M. (2005) The compact city fallacy. *Journal of Planning Education and Research*, vol 25, no 1, pp11–26.

Nordahl, D. (2009) *Public Produce: The New Urban Agriculture*. Washington, DC: Island Press.

Saed (2012) Urban farming: The right to what sort of city? *Capitalism Nature Socialism*, vol 23, no 4, pp1–9.

Schneider, M. and McMichael, P. (2010) Deepening and repairing, the metabolic rift. *Journal of Peasant Studies*, vol 37, no 3, pp461–84.

Sonnino, R. (2009) Feeding the city: Towards a new research and planning agenda. *International Planning Studies*, vol 14, no 4, pp425–435.

Steel, C. (2008) *Hungry City: How Food Shapes Our Lives*. London: Vintage.

Tornaghi, C. (2012a) Public space, urban agriculture and the grassroots creation of new commons: Lessons and challenges for policy makers. In: Viljoen, A. and Wiskerke, J. S. C. (eds) *Sustainable Food Planning: Evolving Theory and Practice*. Wageningen: Wageningen Academic Publishers, pp349–364.

Tornaghi, C. (2012b) Edible public space: Experimenting with a socio-environmentally just urbanism. *Territorio*, vol 60, pp39–43.

Tornaghi, C. (2014a) Critical geography of urban agriculture. *Progress in Human Geography* 38(4), pp. 551–567.

Tornaghi, C. (2014b) *How to Set Up Your Own Urban Agricultural Project with a Socio-Environmental Justice Perspective*. Leeds: University of Leeds.

Tornaghi, C. (2017) Urban agriculture in the food-disabling city: (Re) defining urban food justice, reimagining a politics of empowerment. *Antipode*, vol 49, no 3, pp781–801.

Tornaghi, C. and Certomà, C. (2018) *Urban Gardening as Politics*. London: Routledge.

Tornaghi, C. and Dehaene, M. (2020) The prefigurative power of urban political agroecology: Rethinking the urbanisms of agroecological transitions for food system transformation. *Agroecology and Sustainable Food Systems*, vol 44, no 5, pp594–610.

Tornaghi, C. and Halder, S. (2021/forthcoming) Reconfiguring the intersection between urban food movements and agrarian struggles: Building an urban political agroecology praxis. In: B. Engels, K. Dietz, H. Akram-Lodhi and B. McKay, eds., *The Edward Elgar Handbook of Critical Agrarian Studies*. Cheltenham: Edward Elgar Publishing.

Van Dyck, B., Maughan, N., Vankeerberghen, A. and Visser, M. (2017) Why we need urban agroecology. *Urban Agriculture Magazine*, vol 33, pp56.

Van Dyck, B., Tornaghi, C., Halder, S. von der Haide, E. and Saunders, E. (2018) The making of a strategizing platform: From politicising the food movement in urban contexts to political urban agroecology. In: Tornaghi, C. and Certomà, C. (eds) *Urban Gardening as Politics*. London: Routledge, pp. 167–185.

Viljoen, A. (ed.) (2005) *Continuous Productive Urban Landscapes: Designing Urban Agriculture for Sustainable Cities*. Burlington, MA: Architectural Press.

Viljoen, A. and Wiskerke, J. S. C. (eds) (2012) *Sustainable Food Planning: Evolving Theory and Practice*. Wageningen: Wageningen Academic Publishers.

Wirth, L. (1938) Urbanism as a way of life. *The American Journal of Sociology*, vol 44, no 1, pp1–24.

1

FOOD AS AN URBAN QUESTION, AND THE FOUNDATIONS OF A REPRODUCTIVE, AGROECOLOGICAL URBANISM

C.M. Deh-Tor

Introduction

In this chapter, we aim to present and discuss what we mean by the concept of "agroecological urbanism" (Deh-Tor, 2017). In the introduction to the book, we began by illustrating how encountering agroecology has changed our way of looking at sustainable food planning in a profound way. Here, while delving deeper into the ideas of an agroecological urbanism, we aim to unpack further how we envision a transformative agenda for the sustainable food planning community.

An agroecological urbanism – as a realm of professional practice – does not yet exist. Ours is a call for a dialogue between two sets of discussions and reflections that, until today, still largely operate in separate worlds and are rooted in very different communities of practice. On the one hand are the political agroecology and the food sovereignty movements; these largely represent rural communities engaged in struggles and negotiations at national and transnational levels to shape production and trade conditions of farmers. On the other hand is the urban food policy community that is mostly engaged in debates on urban and regional food strategies, strongly focussed on issues of food access and consumption, such as urban diets and food poverty, but lacks a radical stance on the ecological basis of food production, and the reproduction of life in general. As largely separate movements, they are rooted in very different sets of historical subjectivities and resonate with political positions that have been historically rendered as the conflict between the agrarian and the urban question (McMichael, 2013; Tornaghi and Halder, 2021; Tornaghi and Dehaene, 2020).

While our work is largely positioned in this gap and attempts to build bridges between these communities and movements, in this chapter, we mostly speak to the food planning community. The chapter tries to link up to the unfolding discussion on sustainable urban food planning while trying to break open its agenda: we do this in three ways. First, we aim to inscribe food planning within *a different geography*, moving beyond the city as a self-contained world exploring the complex geometries of planetary urbanisation (Brenner, 2014) and the many concretely existing overlaps between what used to be country and what used to be city (Parham, 2019). Second, we aim to expand and open up *the thematic confines of the urban food agenda*, unpacking the logics of urbanisation that still largely contribute to treating food as an afterthought, after the 'hard' subjects of housing, transport and energy have been taken care of. Third, we aim to challenge *the disciplinary confines* of traditional ways of understanding planning, embracing a view that sees planning as a field in transition, rather than a singular and monolithic disciplinary basis upon which food planning is to be built.

The chapter is organised along four sections. In section 1 we start from the way sustainable food planning has engaged with the urban food question so far, and try to map the boundaries of the terrain that the sustainable food planning agenda was able to conquer within an urban policy context. The aim is to describe its geographical boundaries, the selective character of its political agenda, the main planning approaches followed to implement this agenda, and to begin to illustrate its limits.

In Section 2 we turn to the social reproduction literature as a forceful entry point to rethink the urban food agenda. In particular, we illustrate how the feminist social reproduction literature (Federici, 2004; Mitchell et al., 2004; Bakker and Gill, 2003; Bezanson and Luxton, 2006) has helped us to see the variety of practices that have been residualised and sidelined in a capitalist society that has built the urban question around the question of the reproduction of capital and waged labour (Castells, 1972; Harvey, 1985) rather than bodies and ecologies.

In Section 3 we then move to agroecology as a radical starting point for a new food planning agenda. The clear positioning of political agroecology helps us to map where further articulation is needed in order to creatively imagine and build an urban society that embraces and nurtures the ecological processes that feed life (and us).

In Section 4 we conclude with a call for a heterodox planning practice and try to map some of the voices present in this edited volume within such a heterodox approach.

1. How 'urban' is food planning?

Over the past two decades, since the first publications calling for more attention to the food agenda, sustainable food planning has moved from "being a stranger to the planning field" (Pothukuchi and Kaufman, 2000) to become one of the issues driving the renewal of planning (Morgan, 2013). The reasons for planning communities re-engaging with food are varied and have been described by Morgan and Sonnino as part of the "new food equation" (NFE) (Morgan, 2009; Morgan and Sonnino, 2010). With this term, they refer to the interplay of five profoundly destabilising trends in the capitalist food system that revolve around food, and that could potentially lead to a food regime change (Friedmann, 1987): the sharp rise of staple food prices, increase in food insecurity, the link between food insecurity and national security, the effect of climate change on food production and the growing incidence of land conflicts. Cities, as Morgan and Sonnino (2010) remind us, are at the forefront of the NFE for ecological, demographic and political reasons.

The Food Interest Group (a subsection of the American Planning Association) and the Sustainable Food Planning Group (its counterpart within the Association of European Schools of Planning [AESOP]) can now look back at more than ten years of exchange in research and teaching activities on this topic. The emergence of the field of sustainable food planning went hand in hand with local and regional actors engaging in the drafting of urban food plans, food strategies, and, to some extent, food policies (Ilieva, 2016). Over the last decade, a growing number of cities have installed local food councils (Moragues-Faus and Sonnino, 2019); in turn, this has led to the establishment of networks of cities coming together around the urban food agenda and learning from each other, both within national (i.e., the UK Sustainable Food Cities network) and international networks (i.e., the Milan Urban Food Policy Pact).

We share the excitement for this growing attention. At the same time, however, we feel that the planning community has only started to address food as an *urban* matter of concern. The impact of this renewed attention to food is still rather limited and insufficiently integrated within a broad transformative urban agenda, i.e., an agenda for the city, urbanism and urbanisation more broadly. We believe that the work of planners needs expanding beyond the boundaries within which food has been treated so far and should be understood as an 'urban question', giving it the same weight and centrality that has historically been given to the housing question, mobility, or sanitation in urbanism. The particular gaze we adopt has roots in the literature and debates on 'urban questions' (typically the housing questions), and the ways in which planners' social movements and governing authorities have negotiated issues of social reproduction and collective services in the past century. We first take stock of the ways in which the city has been conceptualised within food planning, then explore the frontiers of an expanded urban agenda.

1.1. Multiple versions of the city in urban food planning

As mentioned above, Morgan and Sonnino (2010) have pointed out that after half a century in which faith in the industrialisation of agriculture made it seem as if the issue of feeding people had been resolved, food has regained centre stage in the international arena. Wiskerke (2015) identifies five urbanisation challenges to which urban food planning has responded:

(1) governance capacity, especially given the new sustainability challenges;
(2) resource use;

(3) growing inequality;
(4) environmental pollution; and
(5) food provisioning for a growing urban population.

Each of these challenges has led to different ways in which the food planning community has engaged with the urban context. Building on these analyses and emerging debates, we can identify the following ways in which the city and food planning intersect.

The city as a distinctive level of governance: the rise of urban food councils and alternative food networks

While in the post-war period food was dealt with largely at the national and supranational level (through trade agreements and price control policies, including subsidies, for example), in the past 30 years, the city has re-emerged as a distinct level of governance for the food system. Moragues-Faus and Sonnino (2019) describe the rise of food policy councils since the establishment of the first council in Knoxville, Tennessee, in 1982. The experience of this growing number of cities has led to the emergence of city-to-city exchanges and city-to-city learning, and is now also available as a rich resource for the empirical analysis of urban food governance (Baker and de Zeeuw, 2015). Food planning in this context is looked at as a specific subject of local and metropolitan governance. The intersections with the urban planning literature are many. Food planning has been analysed as a form of multilevel governance, calling for horizontal and vertical policy integration (Ilieva, 2016). It has been analysed as spaces of hybrid governance and social innovation (Manganelli and Moulaert, 2019), as a new form of regional metropolitan governance, etc. (Wascher et al., 2015). While these new policy arrangements have been celebrated by many, they have also been questioned as symptomatic of a techno-managerial and post-political form of climate governance (Kenis and Lievens, 2014; Swyngedouw, 2010).

The city as a multiscalar territorial entity: reterritorialisation and the rural-urban continuum

The relative neglect of food in the history of urban planning until the mid-1980s is partly explained through the historical physical and mental separation of town and country. Food has been treated as a question of agriculture and constructed as belonging to non-urban territories (Sonnino, 2009; Cabannes and Marocchino, 2018). Many voices have pointed to the region as the preferred geographical entity to rebuild urban-rural linkages (Kneafsey, 2010; Cohen, 2010; Forster and Getz Escudero, 2014; Wiskerke, 2009, 2015). Research has shown great differences in the structure and make-up of the geographical area upon which cities rely (Zasada et al., 2019). However, the concept of urban agriculture has reconnected actors on both sides of the urban-rural divide (Viljoen, 2005) and has actively explored opportunities to integrate food production within the urban context. Designers and planners reimagined the possibilities of reconnecting open spaces along the urban transect (Duany, 2012) into Continuous Productive Urban Landscapes (Viljoen, 2005; Bohn and Viljoen 2010) and discovered the specific opportunities presented by the peri-urban interface (Sieverts, 2003; Parham, 2019).

The city as a contested terrain: urban food movements and the struggle for food justice

While urban food movements across the globe have typically organised around questions of urban hunger and food insecurity, a more organised response to the corporate food regime has united many of these movements around a food justice perspective, striving for equal access to food in cities (Wekerle, 2004; Alkon and Agyeman, 2011). These movements have typically tied into community garden initiatives or community supported agriculture; they have also challenged the enclosure of resources and reclaimed public land as a collective food growing resource (Lyons et al., 2013; Tornaghi, 2012; Certomà and Tornaghi, 2015; Tornaghi and Certomà, 2019). While the food justice movement is typically rooted in race-, gender- and class-based struggles for equality, in their rights-based orientation they are potentially aligned with the more globally oriented and agrarian-based food sovereignty movement. Many urban movements, however, encounter the limits of a neoliberal context (Clendenning et al., 2016). Urban food planning, in general, remains locked up within reformist, at best progressive, policy frameworks that lack a radical and transformative dimension (Holt Giménez and Shattuck, 2011; McClintock, 2014; Tornaghi and Dehaene, 2020).

The city as a dysfunctional ecosystem: mending the urban metabolism

A large section of the food planning literature reflects a general effort to analyse the environmental performance of the urban ecosystem, reducing the extractive use of resources brought in from elsewhere and internalising the negative externalities caused by urban growth. Urban agriculture in particular has been praised for its potential to deliver a broad range of ecosystem services: contributing to the reduction of food waste and the circular use of urban waste streams, the reduction of greenhouse gas emissions, integral urban water management, the use of renewable energy and improved biodiversity (Viljoen and Wiskerke, 2012; Aerts et al., 2016). Many urban food policies are implemented in the context of policy arenas specifically geared at the realisation of urban climate goals. The urban arena and the regional food system have been embraced as plausible scales to analyse and structure the complex relationships within the global food-water-energy nexus (Allouche et al., 2019) and to mend the metabolic rift between town and country (Schneider and McMichael, 2010; Dehaene et al., 2016).

The city as a growing number of mouths to be fed

Many advocates of urban food planning make reference to the UN habitat statistics stating that since 2008 more than half the world's population lives in cities (Morgan and Sonnino, 2010; Wiskerke, 2015; Ilieva, 2016). The growing challenge of feeding the world is increasingly an urban challenge. Carolyne Steel's bestselling book *Hungry City* convincingly showed how urbanisation historically also comes with a growing dependency on processes that largely fall outside of the control of cities (Steel, 2008). At the same time, cities, in particular those of the Global South, have been documented as the habitat of local food initiatives, and forms of self-organisation, contributing to a geography of small- and medium-size cities that still display a significant degree of self-reliance (Mougeot, 2005). Urban agriculture, and even more specifically urban agroecology, has been taken up by the Food and Agriculture Organization of the United Nations (FAO) as an important alternative source of food and contribution to the creation of sustainable livelihoods (Dubbeling et al., 2010; IPES-Food 2018).

Against the background of this landscape, in the next section we aim to contribute towards pushing the boundaries of urban food planning, to discuss how we advocate for food to be fully embraced as an urban question, and to prepare the ground for what we call an agroecological urbanism.

1.2. Food as an urban question: pushing the boundaries of urban food planning

As the field presents itself in more consolidated form, we are also at a point where we see its limits. We see limits in the extent to which the field has questioned current models of urbanisation, the selective uptake of 'urban questions' (i.e., what has been taken up as a matter of urban policy and urban collective arrangements), and in the planning models that have been adopted to address the urban food question.

Urbanisation unbounded: the geography of the urban food question

Food planning by and large still treats the city as a container in which food needs to be retrofitted. This is particularly true for the sometimes uncritical embrace of the full spectrum of urban agriculture initiatives. Urban agriculture has been mostly projected on existing open spaces, fringe spaces, often along infrastructure corridors, on roofs, etc., without fundamentally questioning the land use dynamics that, more often than not, contribute to the further destruction of valuable soils, the contamination of aquifers, the fragmentation of nutrient cycles and the disruption of critical ecosystems. Most food plans accept the division between town and country that places the urban on the consumption side and introduces urban agriculture as a novel exception. Different forms of urban agriculture are typically mapped along the urban transect (Duany, 2012; de Graaf,

2012; Lohrberg et al., 2016) and work under the assumption that the geometry of urban-rural land use dynamics can be adequately captured in terms of centre-periphery relationships.

A more complex understanding of the urban landscape can be found in the framework of the urban food region. This perspective also runs the risk of uncritically embracing the urbanistic consensus that sustainable urbanisation is compact and can be contained within the regional geography of a physiographic basin. While these models may have their role to play, they stand in the way of a food urbanism that embraces the multiscalar geometries of the planetary urban in full (Brenner, 2014). Such exercises may take inspiration from the historical work of early socialist and anarchist thinkers such as Vandervelde (1903), Kautsky (1988) and Kropotkin (1998), who tried to translate the politics of land and labour of industrial urbanisation within a mixed geography of "fields, factories and workshops" (Kropotkin 1998). Equally inspiring are the echoes of these historical reflections in the work of Murray Bookchin (1976) and Colin Ward (1999), or the speculations about Desakota landscapes (McGee, 1991) and Agropolitan Development (Friedmann and Douglass, 1978) in the Global South. All these are speculative exercises that share attention with the agrarian side of the question of urbanisation and resist the extractive, centralistic, cumulative *status quo* of capitalist urbanisation in favour of distributed models in which food production is part and parcel of the urban landscape. In our quest for a radical rethinking of food planning, our first key message is a call for overcoming the artificial and capitalism-driven geographical and conceptual separation of what is thought of as 'agricultural lands' from what is conceived as 'urbanisation'.

The selective uptake of urban questions: the food question as an afterthought

Food planning is in part a response to the selective and limited ways in which food has been treated as an urban question over the past 100 years. Urban questions have historically been concerned with social reproduction under urbanisation (Castells, 1972) and the exacerbated relationships of interdependence that urban life comes with. While answers to the urban social reproduction crisis have not always been necessarily urban, the interplay between social movements and business interests has posed considerable pressures on the urban political agenda to provide an answer. Affordable housing and cheap food were equally essential to the reproduction of industrial waged labour; however, their different historical treatment, and the different ways in which they have been commodified, is striking. Transport and commuting have been key to keeping house prices low and the proletariat out of the city, but public housing provision would always remain central to the urban agenda (Polasky, 2010). The urban food question could have been resolved through protection of urban and peri-urban agricultural land, an investment in logistics, municipal slaughterhouses, or the appropriate provision of wholesale markets, while in fact they have been used only to facilitate the import of food from elsewhere (Cronon, 1992; Steel, 2008) and to enable the massive conversion of agricultural land to urban land use. While housing, transport and sanitation have made it into the main chapters of the urban planning textbooks, food was for the footnotes, only to be rediscovered lately as a novelty in the field.

While cities have begun to subscribe to (slightly) more ambitious goals, such as those formulated within the Milan Food Policy pact or by the FAO, the translation of these goals in actual policies remains attached to those policy areas that have historically landed on the urban side (FAO 2018, 2019). Urban food policies tend to focus on consumption and consumer behaviour, insufficiently questioning the extreme state of food commodification within cities. Strategies to decommodify food remain attached to welfare measures typically associated with food poverty (i.e., food vouchers), compensating for the failures of a commodified urban foodscape rather than aiming for its fundamental transformation. In order to put food on the urban agenda, food policies have been tactically linked to urban climate governance, public health policies, food poverty measures, or waste management. These tactical alignments do not necessarily break open the structural barriers that exist between divided policy silos, leaving the potential of a food perspective to contribute to policy integration largely untapped.

Our second key message for a renewed planning approach to food is the call to fully embrace the production of food (not only its consumption) as an 'urban question', a question that needs to be responded to locally. In order for food to surge to the same level as housing or transport in the planning agenda, it is necessary that land and soils enter the political agenda and reshape the politics of resource management. This means that land for food production and the preservation of healthy soils do not come secondary to (and hence sacrificed for) the expansion of road infrastructure, housing development, or new commercial areas. While it is obvious that no urban context alone is expected to meet its food needs, the surging of food to a fully articulated urban question will require new fields of work to be defined and articulated in ways that substantially redefine the hardwiring and the software of our urban environments. This is in terms of land use (i.e., proactive policies for enhanced land protection in the urban and peri-urban, programmes for the use of existing public land and plans for reconversion of land to agricultural use), infrastructure (i.e., build or rebuild the infrastructure for food production, including use and redirection of organic waste streams, legal protection of soils, programmes for soil regeneration, waterways sanitation and restoration for agricultural use, creation of municipal free-seeds banks, etc.) and logistics (institution of municipal storage spaces).

The hegemony of established planning ideas: a food blind planning guild

Food might be innovative for planning, but is planning sufficiently innovative to address the urban food question? Some humility is in place here. As Ilieva puts it, "planning might be part of the problem":

> For many years, urban plans have labelled periurban lands around cities as 'awaiting development' and hatched them as blank space, disregarding the great diversity of rural infrastructures and landscapes that distinguish one periurban area from the other. Urbanization proceeds regardless of these diversities and thus has had a detrimental impact on many peri urban farms and rural heritage sites, particularly in European urban regions.
>
> *(Ilieva, 2016, 80)*

The great challenge is to see the historical blindness of planning to food. Changing that does not only require a new song but also touches the epistemological foundations of planning. This includes the conceptual hierarchy that is embedded in the field, that thinks from the centre and disqualifies the rest as periphery (Angelo and Wachsmuth, 2015), that gives disproportionate weight to questions of real estate, housing, transport, etc., the subjects that drive the urban growth machine, that operates through legal frameworks and technical instruments that reproduce this conceptual hierarchy.

Our third key message for the renewal of food planning is that this is not possible without a deep rethinking of the field of planning itself. The very notion of planning, as deeply anthropocentric, rooted in instrumental rationality, the domination of nature, and historically complicit with the power geometries of a colonial world order, may even be the wrong label to imagine a food-inclusive, resource-conserving and regenerative urban world. The wager to fully think food as an urban question, however, may be a concrete and tangible starting point to imagine a future for 'the field formerly called planning', including the delivery of its emancipatory aspirations.

2. Urbanism, food and social reproduction

Planning's historical blindness to food can be better understood (and challenged) in all its consequences if we look at the history of planning and urbanisation from the perspective of social reproduction. Feminist social reproduction scholars have provided fundamental insights into the social and material reproduction of societies, including giving birth, care work and food provisioning; these have been historically ensured over time.[1] The rise of urban planning in the last century was met with an ongoing critique of the links between capitalist industrialisation and processes of urbanisation. Scholars in the 1960s and 1970s exposed how urbanisms were ideologically driven and value-laden practices organising urban space and life, functional to the dominant economic, political and social order. Colonial, capitalist (and patriarchal) values shaped the articulation of urban space needed to ensure the maintenance and thriving of capitalist societies, processes of accumulation and resource extraction. Critical urban scholars, such as Henri Lefebvre, Manuel Castells and David Harvey, have pointed out, for example, how the expansion of the built environment and the destruction of fertile lands were essentially linked to the circuits of capital and the search of its ongoing expansion (real estate speculation, capital spatial fix). They also describe how state-led provisions of collective services, such as education and social housing (elements of social reproduction), were functional toward enabling a workforce (once based in the countryside as at least partially self-sufficient farmers) to be concentrated around factories, and how urban ways of life had a growing environmental and social effect on a planetary scale. These effects included the ongoing depletion or pollution of natural resources, the development of tourist resorts on virgin lands and the subjugation of agriculture to commodity markets. All these were well beyond the sphere of the urban.

Feminist scholars have provided fundamental critiques and much-needed integrations to these theories, shaping a 'social reproduction' perspective. They observed that a fundamental element in the ongoing renewal of capital relationships was not only the appropriation of means of production on behalf of the capitalist (as most Marxist literature has unpacked), but also the unpaid and unrecognised labour necessary for the reproduction of the workers, which were often the responsibility of women (Dalla Costa and James, 1975; Federici, 2004, 2012, 2018). These included, alongside the obvious biological reproduction of life (giving birth), all the domestic and care work needed for the day-to-day reproduction of able workers: providing for the satisfaction of sexual needs, the preparation of food, the washing and sewing of clothes, care for the children and the elderly, the education and socialisation to cultural norms and social rules (Fraser, 2016, p.23; Jacka, 2017). These activities were not only necessary for the reproduction of society and life in general, but were often a direct outcome of oppressive social arrangements emanating from patriarchal values.

Some of the feminist scholars active in this field had begun to critique how the urban design and architecture informing new urban and suburban development were often implicated in the reproduction of isolating living conditions, which rendered impossible the socialisation of these tasks (Hayden, 1982). Socialist and anarchist material feminist groups, especially in the period between the end of the 19th century and the beginning of the 20th century, engaged in the conceptualisation and development of concrete alternatives for the management of these tasks in common, including experiments in urban planning and development. At a time when the male-dominated professions tried to render housework more efficient through the technological innovations of modern utility systems and imagined apartments equipped with housekeeping machines and kitchens built into each minimum dwelling, the imaginaries of these groups were centred around public/collective kitchens, kitchenless houses, co-operative housekeeping, communal living and community-led urban agriculture (see, for example, the work of Melusina Fay Peirce, Charlotte Perkins Gilman, Mary Livermore, Ethel Puffer Howes, Charles Fourier, Ebenezer Howard, among others).

However, while pointing out that social reproduction has been undervalued and under-investigated, they also point out how it remained secondary and subservient to capitalist production. It is important for our argument to focus on the conditions of the provision of food.

First, it is important to remember that capitalism had deep roots in agriculture in the 15th century, with the expropriation of land and common lands to farmers, what is called 'primitive accumulation', and the rise of international trades of goods, and later slaves. The creation of masses of dispossessed, destitute people was the pre-condition for the rise of industrialisation and rapid urbanisation.

Under conditions of commodification of labour, workers become

> dependent on market for the items they once produced at home (or obtained through the informal economy). … [W]ithout a wage they cannot obtain crucial subsistence goods. It is because of this cycle of dependence that the market actually comes to dominate social reproduction in general.
>
> *(Ferguson, 1998:4, in Bakker and Gill, p.21)*

Alongside the commodification of labour, family-based food production/gardening/animal rearing, that had been common practice and survived in many urban contexts, has progressively been residualised and food provision treated as a commodity.

The agro-industrialisation of food, supported by the green revolution, as well as various forms of dumping of "cheap food from nowhere" (Bové, et al. 2002), have naturalised the abstraction of agriculture from its ecological and cultural foundations (McMichael, 2003, p.173). Over time we have become used to think of the urban being on the side of consumption and rural on the side of production. Alongside the furthering of primitive accumulation, the enclosure of common lands, the urbanisation of residualised fields and the set of hygiene standards, many food-provisioning practices have been rendered impossible and their scope and diversity severely reduced. People have grown disconnected from the ecological basis of food production, its seasonality and the importance of returning organic waste (i.e., kitchen waste) to the land to feed the soil. They have lost not only fundamental knowledge to understand our mutual interdependence with other life organisms (what Schneider and McMichael have discussed as 'epistemic rift') and the ability to provide for their survival, but they have also lost the very possibility of reproducing such knowledge: they have lost the possibility of practising it (Schneider and McMichael, 2010; Tornaghi, 2017).

In the consolidated urban environments of the Global North, public and private spaces are not generally designed to accommodate food self-provisioning communities, nor do they offer the infrastructure needed to serve urban and peri-urban smallholders. Land values and land tenure regimes devalue food production and cater for speculative and profit-oriented approaches to land 'development'. Organic waste management is driven by short-sighted cost-saving approaches, rather than soil-nurturing and nutrient-recovery practices. A social reproduction perspective enables us to see the specific biopolitical character of those forms of urbanisation that have institutionalised and reproduced urban ways of living that assume the endless provisioning of cheap food 'from nowhere'. This food is produced unsustainably by underpaid farm workers, wasted in disproportionate amounts and often processed in unhealthy ready-meals because people increasingly do not have the available time or skills to cook.

Social reproduction scholars have indeed pointed out how the progressive absorption of women into the labour market and the ongoing retreat of the state from social provision of welfare have coincided with a re-privatisation of social reproduction (Mitchell et al., 2004) by either externalising care work to paid workers or a re-burdening of women with care work, often in addition to full-time waged work. To heal the knowledge and epistemic rift, we do not only need to promote a reconnection of farmers with urban dwellers but also need to create time and space for those practices needed to reproduce (embodying) lost knowledge.

Sustainable food planning, in its attempt to transform the food system, therefore, needs to understand and challenge the profound interconnection between capitalist and neoliberal values and practices on the one hand, and the arrangements for social reproduction ingrained in processes of urbanisation on the other. These efforts should not stop at the critical deconstruction of the urbanisms of capital. We see it as a call to imagine alternative urbanisms that decouple urbanism and capital and seek to imagine forms of urbanisation that value the social reproduction of life in general, and of food in particular, differently (Tornaghi and Dehaene, 2020). This is a tall order given the intimate historical relationship between urbanisation and capitalism that has defined the ways cities have been structured since the rise of industrial capitalism. Our call for an agroecological urbanism (which we will work out further in the next section), in order to transform our food system, takes inspiration from political agroecology, as a movement that offers concrete alternative value systems, social arrangements and ecological practices from where to imagine a post-capitalist world.

3. Agroecology and the re-articulation of the urban food agenda: imagining an agroecological urbanism

> Agroecology – in our view – is not just an agricultural method: it is a 'package' of value-based practices which are explicitly addressing social and environmental justice, are culturally sensitive, non-extractive, resource conserving, and rooted in non-hierarchical and inclusive pedagogical and educational models that shape the way food is produced and socialised across communities and generations. Agroecosystems, while specific to each geographical context, share a number of ecological and social features including 'socio-cultural institutions regulated by strong values and collective forms of social organisation for resource access, benefits sharing, value systems'. The principles and practice of agroecology, centred around multi-species solidarities, biodiversity and environmental stewardship, have been extensively noted for their ability to conceive of and deliver *alternative ways of producing food*. Agroecology is also being strongly mobilised as *a political tool*. Its strong links with the international food sovereignty movement,

> and its inclination to action-oriented, transdisciplinary and participatory processes has led to defining it simultaneously as a science, a movement and a practice. Political agroecology and urban political agroecology are taking shape at the crossroads between scholar activism and urban movements, although its full political potential is yet to be metabolised.
>
> *(Deh-Tor 2017)*

Political agroecology has been the framework within which we have embraced the lessons learnt from feminist social reproduction scholars and began to push the reimagining of the urban food planning agenda. Below we explore how agroecology may push the geographical boundaries of the discipline, the political agenda and the disciplinary scope of sustainable food planning.

Agroecology and the biopolitics of the productive city

Thinking the urban through an agroecological lens is more than a call to rebuild urban-rural linkages within the existing geography of capitalist urbanisation. Rather it challenges the mechanisms that reaffirm their separation, with production on one side and consumption on the other, with agrarian versus urban questions, but also with nature on the outside and technology, culture and artifice on the inside. An agroecological urbanism is a political ecological project that seeks to rethink the 'nature of cities' (Heynen and Swyngedouw, 2005) and tries to imagine urban ways of life that relate concretely to, make space for and are centred on the ecological basis of food production.

As reproductive work is enmeshed with production and cannot really be thought of separately, the reshaping of an urban food agenda cannot just be done by plugging agroecology and food into the interstices of the system. Recall that an urbanism is made up of the collective arrangements that a society makes to organise the collective support and care in the context of exacerbated conditions of interdependence, the agroecological urbanism we imagine joins the political path of agroecology and builds on the quest for nature-inclusive forms of agriculture in order to imagine an urbanism that thinks social interdependence and more-than-human interspecies solidarities together (Haraway, 2016; Puig de la Bellacasa, 2017).

Rethinking the biopolitics of the city in an agroecological perspective means to shape an urbanism that re-operationalises the links between urban green spaces and farming fields in the peri-urban fringe, between urban dwellers and food transformers, and between organic waste collectors and territorial food hubs. Within this reflection we see a specific role for the peri-urban context where the processes that work towards the separation of the agrarian and urban question can be seen in action. It is a context in which 'urban', 'rural', 'no longer rural' and 'already urban' communities are co-present and where we see possibilities to imagine the construction of an agroecological urban subject. It is a reality where the skills, the farmers' operating infrastructure and the differentiated landscapes that historically enabled regenerative farming practices are often still in place in residualised or fragmented form. In the midst of such landscapes in transition, an alternative order could be established, starting from the reappropriation of such residualised landscapes, taking back the nutrient scapes and valuable soils, mobilising remaining skills and activating remaining proximity relations between rural and urban fragments.

An equally important role in the re-organisations of these relations can be played by the neighbourhood 'political community kitchen', a place that can act at the interlink between communities (peri-urban food producers, food processors, community caterers, urban gardening and reskilling groups), food-producing activities and territorialities, recasting the neighbourhood scale as central in the building of resourceful communities.

Agroecology and the urban politics of ecological resources

The agenda for an agroecological urbanism is one that has to be literally built with constituencies that belong to worlds that have been separated by the urbanisation of capital. While the agenda of the political agroecology movement has been built in urban exile, with the urban often seen as a direct threat and driver of the logics of dispossession that movements such as the Via Campesina have been fighting, we see agroecology as a comprehensive agenda that provides the key principles upon which a socially inclusive, ecologically sustainable and resourceful urbanism could be built, that is a way of organising the urban that would be putting the care for its social reproduction central.

Agroecology as a science, a movement and a practice (Wezel et al., 2009) is a rich and contextually diverse repertoire of very concrete ideas that can be mobilised by communities 'in place' and around which new collective arrangements can be built. The agenda of the agroecology movement has been typically defined in terms of the contrasting logics of agroecosystems and conventional farming. Gliessman (2007, pp.8–16) builds on this to identify seven areas in which agroecology could play a curative and transformative role: soil degradation, overuse of water and damage to hydrological systems, pollution of the environment, dependence on external inputs, loss of genetic diversity, loss of local control over agricultural production and global inequality.

If we understand the close connection between the process of urbanisation and the expansion of large-scale, extractive forms of conventional farming, we immediately feel how this list could equally lead to the construction of a transformative urban agenda. Such an agenda would:

(i) link land use to soil care and interrupt the logics of substitution that make it possible to bring food to the city rather than produce it locally;
(ii) give a central role for agroecological farmers as stewards of the watersheds that urban systems are typically part of;

(iii) team up with agroecology not only to reduce the polluting effects of industrial agriculture but also to engage in the rebuilding of topsoils on urban damaged lands, including the use of organic material from urban waste streams;
(iv) join forces with nature conservationists and build rich gradients and connections between the intrinsic biodiversity of conservation (or rewilding projects) and the functional biological diversity of agroecological farming;
(v) adopt the principles of resource sovereignty as a means to reinforce both farmers' autonomy as well as a tool to rebuild some local control over food security;
(vi) challenge ethnocentric and exclusive constructions of territorial sovereignty in favour of place-based solidarities that embrace the full diversity of the urban demography.

Agroecology and the politics of (planning) knowledge

Political agroecology has convincingly made the link between a systemic and systematic analysis of agroecosystems and food systems. That relationship is extremely complex: the transformation of the system requires not only a radical transformation of the way processes of urbanisations are organised but also a partial renewal of the type of planning that is needed. To successfully nurture and build an agroecological urbanism, new concerns need to be put at the centre of the planning profession and new skills need to be taught in planning schools. While sustainable food planning has been an innovative force in planning education, the call for an agroecological urbanism is a call to move out of planning schools into the world of agroecology and back, exposing a new generation of planning students and professionals to matters of concern that were not included in the traditional imaginaries of an urban-based community of practice.

We see great potential in forms of advocacy planning that literally translate the principles of agroecological food production, in all their local and cultural diversity, into an agenda for the transformation of the urban landscape from a food disabling into a food enabling one. Such a food planning project is not just a technical endeavour but requires political pedagogies that make new positions available for the food-alienated citizens, subjects of the urbanisms of capital. We imagine concrete advocacy working with food growing communities, identifying the building blocks of an 'agroecological mode of life', and finding the forms and collective articulations for a society that places food practices central in caring for its social reproduction.

While such a call may sound overly abstract, the beauty of focussing on agroecological imaginaries and principles is that such work could find direct and concrete roots in concrete communities and places, in the existing efforts of the very people who today try to create the circumstances to care for the soil, to care for their plants, to care for their communities and face very specific, contextual, daily challenges to do so. It is from these cumulated efforts of political agroecology communities that the force of an urban agroecological transformation could be derived. An agroecological urbanism, then, is more than an agenda: it is also a quest for the platforms and spaces through which such an agenda can be built and supported. We take inspiration from the efforts to root the renewal of urban planning within innovative and radical forms of urban governance, such as radical municipalist projects or communities of food commoners. Where the national planning and supra-national planning arenas continue to cater to the vested interests of the agro-industrial complex, we see potential and precedents to bring a radical municipalist agenda (Russell, 2019) and the struggles of political agroecology together. The first initiatives to build networks of agroecological cities point towards such an agenda (López et al., 2017).

4. Resourcing an agroecological urbanism within a heterodox community of practice

> What if solidarity, mutual learning, interspecies (more than human) exchanges, environmental stewardship, food sovereignty and people's resourcefulness were the principles of a new paradigm for urbanisation? How would urban design, property regimes, food provision, collective services, and the whole ensemble of planning and socio-technical arrangements change, if they were informed by urban agroecology? How can we begin to radically transform the food disabling urban landscapes that have systematically displaced food production, recovering both historical food growing practices and imagining new urban arrangements?
>
> *(Deh-Tor, 2017)*

The incipit above illustrates the kind of questions that, in 2017, led us to call for an international forum for an agroecological urbanism. We were looking for allies within the food planning community to embrace the principles of agroecology, and the tools of political agroecology to join forces with social movements and food activists to build a constructive vision for an agroecological urbanism. The forum we imagined does not exist in consolidated form but the project moved ahead on different fronts. The conversation with the AESOP sustainable food conference continued with sustained attention for the intersection of agroecology and sustainable food planning in the 2019 conference in Madrid, hosted by Marian Simon Rojo. We (the authors of this chapter) received funding from JPI Urban Europe and the Belmont Forum to explore pathways for resourcing marginalised peri-urban agroecological farmers and for the conceptualisation of concrete building blocks for an agroecological urbanism, with partners in Riga, Brussels, Rosario and London (Tornaghi and Dehaene, 2020). This edited volume is also an attempt to bring together pieces of the urban agroecology puzzle.

The voices in this book are quite diverse. A few chapters squarely belong to the community of food planners and are authored by people who identify themselves in those terms. For many, the question of planning, let alone urbanism, is by no means the starting point of their engagement. One of the great merits of the food planning community, however, is that it has contributed to the encounter of heterodox voices and thereby to an expanded understanding of planning. In a moment of growing production in the field of sustainable food planning, what is becoming increasingly clear is that there is wide divergence regarding the kind of planning that is being promoted, and the specific role assigned to cities and the urban policy arena. We believe that it is important not to lock up the urban food question within the confines of disciplinary boundaries, but rather to link up to the diverse movements leading food systems innovation.

The selection of contributions in this book reflects this diversity, without any ambition of being comprehensive. Together they form a grid of entries into a field. Most importantly, the various contributions show different trajectories through which an expanded and re-politicised urban food agenda can be built and identify some of the communities that could be mobilised or are already mobilised. The authors in this book share the belief that to bring about an agro-ecological urbanism, a project so fundamentally at odds with capitalist urbanisation and the dominant food systems in place, the road ahead is a political one that requires clear positions regarding the geography and territoriality of the food planning agenda, the value positions that define what is appreciated and depreciated, reproduced or discarded and the subjectivities and terms of engagement through which such a transformative agenda could be moved forward.

The book includes several chapters that discuss the methods of mobilisation and engagement that could bring about situations in which humans and non-humans could enter into more virtuous relationships and take up roles for which they now lack the resources, the skills, or the imagination. The book includes accounts of concrete participative forms of action research that has sought to include agroecological farmers within debates on food systems transformation and sustainable food planning (in particular Chapters 2, 7 and 8). Several contributions reflect on the counter-hegemonic strategies necessary to bring about an agroecological urbanism. These could come in the form of the politics of the commons (Chapter 3), could take inspiration from the solidarity economies built by the Chiapas in Mexico (Chapter 9), or may require global solidarities between precarious communities in the North and peasant movements in the Global South (Chapters 8 and 10). Several chapters start from the need to challenge and interrupt the mechanisms that reproduce the problems of our current food system, its inequalities and logics of dispossession (Chapter 10), the pervasive contamination of soils in general and urban soils in particular (Chapter 6), the loss of skills and the extensive alienation of urban subjects (Chapters 2 and 7) and the reproduction of colonial and extractive relationships with capitalistic urban societies (Chapter 3). Some chapters build on specific practices that hold the promise of engendering new agroecological relationships within our urban landscape (Chapters 4 and 5). Chapter 5 explores the way in which the introduction of food-producing perennials and principles of agroforestry could transform the way we look at open space resources and the landscape ecology of the urbanised landscape. Chapter 4 discusses the reappropriation of fragmented lands on the peri-urban fringe by agroecological smallholders.

Overall, we see the authors of these chapters as travel companions, and we hope the book will inspire other practitioners, scholars and scholar-activists to join in reimagining and building an agroecological urbanism.

Note

1 Given the space constraints, we focus this discussion around the last 150 years, although Silvia Federici (2004) and Harriet Friedmann (1987) have provided detailed and enlightening overviews of the social and international relationships that impacted on both food regimes and women's lives, covering a history dating back to the 15th century or earlier.

References

Aerts, R., Dewaelheyns, V. and Achten, W. M. J (2016) 'Potential ecosystem services of urban agriculture: A review'. *PeerJ Preprints*, vol 4, ppe2286v1.

Alkon, A. H. and Agyeman, J. (eds) (2011) *Cultivating Food Justice: Race, Class, and Sustainability*. Cambridge, MA: MIT Press.

Allouche, J., Middleton, C. and Gyawali, D. (2019) *The Water–Food–Energy Nexus: Power, Politics, and Justice (Pathways to Sustainability)*. New York: Routledge.

Angelo, H. and Wachsmuth, D. (2015) 'Urbanizing urban political ecology: A critique of methodological cityism'. *International Journal of Urban and Regional Research*, vol 39, no 1, pp16–27.

Baker, L. and de Zeeuw, H. (2015) 'Urban food systems'. In: H. de Zeeuw & P. Drechsel (eds), *Cities and Agriculture. Developing Resilient Urban Food Systems*. London: Routledge, pp26–55.

Bakker, I. and Gill, S. (eds) (2003) *Power, Production and Social Reproduction*. London: Palgrave Macmillan.

Bezanson, K. and Luxton, M. (2006) *Social Reproduction. Feminist Political Economy Challenges Neo-Liberalism*. Montreal, Canada: McGill-Queen's University Press.

Bohn, K. and Viljoen, A. (2010) 'The edible city: Envisioning the continuous productive urban landscape (CPUL)'. *Field Journal*, vol 4, no 1, pp149–161.

Bookchin, M. (1976) 'Radical agriculture'. In: R. Merrill (ed.), *Radical Agriculture*. New York: New York UP, pp3–13.

Bové, J., Dufour, F. and Luneau, G. (2002) *The World is Not for Sale: Farmers against Junk Food*. London: Verso.

Brenner, N. (2014) *Implosions/Explosions. Towards a Study of Planetary Urbanization*. Berlin: Jovis.

Cabannes, Y. and Marocchino, C. (eds) (2018) *Integrating Food into Urban Planning*. London: UCL Press.

Castells, M. (1972) *La Question Urbaine*. Paris: Maspero.

Certomà, C. and Tornaghi, C. (2015) 'Editorial. Political gardening. Transforming cities and political agency'. *Local Environment: The International Journal of Justice and Sustainability*, vol 20, no 10, pp1123–1131.

Clendenning, J., Dressler, W. H. and Richards, C. (2016) 'Food justice or food sovereignty? Understanding the rise of urban food movements in the USA'. *Agriculture and Human Values*, vol 33, no 1, pp165–177.

Cohen, N. (2010) 'How great cities are fed revisited: Ten municipal policies to support the New York city foodshed'. *Fordham Environmental Law Review*, vol 22, no 3, pp691–710.

Cronon, W. (1992) *Nature's Metropolis: Chicago and the Great West*. New York: WW Norton & Company.

Dalla Costa, M. and James, S. (eds) (1975) *The Power of Women and the Subversion of the Community*. Bristol: Falling Wall Press.

de Graaf, P. A. (2012) 'Room for urban agriculture in Rotterdam: Defining the spatial opportunities for urban agriculture within the industrialised city'. In: A. Viljoen and J. S. C. Wiskerke (eds), *Sustainable Food Planning: Evolving Theory and Practice*. Wageningen: Wageningen Academic Publishers, pp533–546.

Deh-Tor, C. M. (2017) 'From agriculture in the city to an agroecological urbanism: The transformative pathway of urban (political) agroecology'. *Urban Agriculture Magazine*, vol 33, pp8–10.

Dehaene, M., Tornaghi, C. and Sage, C. (2016) 'Mending the metabolic rift – Placing the 'urban' in urban agriculture'. In: F. Lohrberg, L. Lička, L. Scazzosi and A. Timpe (eds), *Urban agriculture Europe*. Berlin: Jovis, pp174–177.

Duany, A. (2012) *Garden Cities: Theory and Practice of Agrarian Urbanism*. London: The Prince's Foundation for the Built Environment.

Dubbeling, M., de Zeeuw, H. and van Veenhuizen, R. (eds) (2010) *Cities, Poverty and Food: Multi-Stakeholder Policy and Planning in Urban Agriculture*. Bourton on Dunsmore: Ruaf Foundation and Default Book Series.

FAO (2018) *The Role of Cities in the Transformation of Food Systems: Sharing Lessons from Milan Pact Cities*. Available from http://www.milanurbanfoodpolicypact.org/wp-content/uploads/2018/10/CA0912EN.pdf.

FAO (2019) *FAO Framework for the Urban Food Agenda. Leveraging Sub-national and Local Government Action to Ensure Sustainable Food Systems and Improved Nutrition*. Available from http://www.fao.org/3/ca3151en/CA3151EN.pdf.

Federici, S. (2004) *Caliban and the Witch: Women, the Body and Primitive Accumulation*. Brooklyn: Autonomedia.

Federici, S. (2012) *Revolution at Point Zero: Housework, Reproduction and Feminist Struggle*. Oakland, CA: PM Press.

Federici, S. (2018) *Re-enchanting the World. Feminism and the Politics of the Commons*. Oakland, CA: PM Press.

Ferguson, S. (1998) 'Building on the strengths of the socialist feminist tradition'. *New Politics*, vol 7, no 2, pp89–100.

Forster, T. and Getz Escudero, A. (2014) *City Regions as Landscapes for People, Food and Nature*. Washington, DC: EcoAgriculture Partners.

Fraser, N. (2016) 'Contradictions of capital and care'. *New Left Review*, vol 100, no 99, p117.

Friedmann, H. (1987) 'International regimes of food and agriculture since 1870'. In: T. Shanin (ed.), *Peasants and Peasant Societies*. Oxford: Basil Blackwell, pp258–276.

Friedmann, J. and Douglass, M. (1978) 'Agropolitan development: Towards a new strategy for regional planning in Asia'. In: F-C. Lo and K. Salih, (eds), *Growth Pole Strategy and Regional Development Policy*. Oxford: Pergamon, pp163–192.

Gliessman, S. R. (2007) *Agroecology: The Ecology of Sustainable Food Systems*, Second Edition. Boca Raton, FL: CRC Press.

Haraway, D. (2016) *Staying with the Trouble. Making Kin in the Chthulucene*. Durham: Duke University Press.

Harvey, D. (1985) *The Urbanization of Capital*. Baltimore, MD: Johns Hopkins.

Hayden, D. (1982) *The Grand Domestic Revolution: A History of Feminist Designs for American Homes, Neighborhoods, and Cities*. Cambridge, MA: MIT Press.

Heynen, N., Kaika, M. and Swyngedouw, M. (eds) (2005) *In the Nature of Cities. Urban Political Ecology and the Politics of Urban Metabolism*. London: Routledge.

Holt Giménez, E. and Shattuck, A. (2011) 'Food crises, food regimes and food movements: Rumblings of reform or tides of transformation?' *The Journal of Peasant Studies*, vol 38, no 1, pp109–144.

Ilieva, R. (2016) *Urban Food Planning. Seeds for Transition in the Global North*. London: Routledge.

IPES-Food (2018) *Breaking Away from Industrial Food and Farming Systems: Seven Case Studies of Agroecological Transition*. Available from http://www.ipes-food.org/_img/upload/files/CS2_web.pdf.

Jacka, T. (2017) 'Translocal family reproduction and agrarian change in China: A new analytical framework'. *Journal of Peasant Studies*, vol 45, no 7, pp1341–1359.

Kaika, M. and Swyngedouw, E. (2011) 'The urbanization of nature: Great promises, impasse, and new beginnings'. In G. Bridge and S. Watson (eds), *The New Blackwell Companion to the City*. Malden, MA: Wiley-Blackwell, pp96–107.

Kautsky, K. (1988) *The Agrarian Question, Two Volumes*. Translated by Pete Burgess. London: Zwan Publications.

Kenis, A. and Lievens, M. (2014) 'Searching for 'the political' in environmental politics'. *Environmental Politics*, vol 23, no 4, pp531–548.

Kneafsey, M. (2010) 'The region in food—important or irrelevant?'. *Cambridge Journal of Regions, Economy and Society*, vol 3, no 2, pp177–190.

Kropotkin, P. (1998) *Fields, Factories and Workshops Tomorrow* (edited by Colin Ward). London: Freedom Press.

Lohrberg, F., Lička, L., Scazzosi, L. and Timpe, A. (eds) (2016) *Urban Agriculture Europe*. Berlin: Jovis.

López, D., Alonso, N., Herrera, P. M., Mérida, J., Pérez, J. M. (2017) 'Cities for agroecology' networks in Europe and Spain'. *Urban Agriculture Magazine*, vol 33, pp55–57.

Lyons, K., Richards, C., Lotus Desfours, L. and Amati, M. (2013) 'Food in the city: Urban food movements and the (re)- imagining of urban spaces'. *Australian Planner*, vol 50, no 2, pp157–163.

Manganelli, A. and Moulaert, F. (2019) 'Scaling out access to land for urban agriculture. Governance hybridities in the Brussels-capital region'. *Land Use Policy*, vol 82, pp391–400.

McClintock, N. (2014) 'Radical, reformist, and garden-variety neoliberal: Coming to terms with urban agriculture's contradictions'. *Local Environment: The International Journal of Justice and Sustainability*, vol 19, no 2, pp147–171.

McGee, T. G. (1991) 'The emergence of desakota regions in Asia: Expanding a hypothesis'. In: N. Ginsburg, B. Koppel and T. G. McGee, (eds), *The Extended Metropolis: Settlement Transition in Asia*, Honolulu, HI: University of Hawaii Press, pp3–25.

McMichael, P. (2003) 'Food security and social reproduction: Issues and contradictions'. In: I. Bakker and S. Gill (eds) *Power, Production and Social Reproduction: Human In/Security in the Global Political Economy*. London: Palgrave Macmillan, pp169–189.

McMichael, P. (2013) *Food Regimes and Agrarian Questions, Agrarian Change & Peasant Studies*. Halifax: Fernwood.

Mitchell, K., Marston, S. A. and Katz, C. (2004) *Life's Work: Geographies of Social Reproduction*. Oxford: Blackwell.
Moragues-Faus, A. and Sonnino, R. (2019) 'Re-assembling sustainable food cities: An exploration of translocal governance and its multiple agencies'. *Urban Studies*, vol 56, no 4, pp778–794.
Morgan, K. (2009) 'Feeding the city: The challenge of urban food planning'. *International Planning Studies*, vol 14, no 4, pp429–436.
Morgan, K. (2013) 'The rise of urban food planning'. *International Planning Studies* vol 18, no 1, pp1–4.
Morgan, K. and Sonnino, R. (2010) 'The urban foodscape: World cities and the new food equation'. *Cambridge Journal of Regions, Economy and Society*, vol 3, no 2, pp209–224.
Mougeot, L. J. A. (2005) *Agropolis: The Social, Political and Environmental Dimensions of Urban Agriculture*. London: Earthscan.
Parham, S. (2019) 'Foodscape and food urbanism in Europe: The urban-rural interface'. In: E. Gottero (ed.), *Agrourbanism: Tools for Governance and Planning of Agrarian Landscape*. Cham: Springer, pp109–129. doi: 10.1007/978-3-319-95576-6_8.
Polasky, J. L. (2010) *Reforming Urban Labor. Routes to the City, Roots in the Country*. Ithaca, NY: Cornell University Press.
Pothukuchi, K. and Kaufman, J. (2000) 'The food system: A stranger to the planning field'. *Journal of the American Planning Association*, vol 66, no 2, pp113–124.
Puig de la Bellacasa, M. (2017) *Matters of Care: Speculative Ethics in More Than Human Worlds*. Minneapolis, MN: University of Minnesota Press.
Russell, B. (2019) 'Beyond the local trap: New municipalism and the rise of the fearless cities'. *Antipode*, vol 51, no 3, pp989–1010.
Schneider, M. and McMichael, P. (2010) 'Deepening and repairing, the metabolic rift'. *Journal of Peasant Studies*, vol 37, no 3, pp461–84.
Sieverts, T. (2003) *Cities without Cities: An Interpretation of the Zwischenstadt*. London: Routledge.
Sonnino, R. (2009) 'Feeding the city: Towards a new research and planning agenda'. *International Planning Studies*, vol 14, no 4, pp425–435.
Steel, C. (2008) *Hungry City: How Food Shapes Our Lives*. London: Random House.
Swyngedouw, E. (2010) 'Trouble with nature: 'Ecology as the new opium for the masses''. In: P. Healey, and J. Hillier (eds), *The Ashgate Research Companion to Planning Theory: Conceptual Challenges for Spatial Planning*. London: Ashgate, pp299–320.
Tornaghi, C. (2012) 'Edible public space: Experimenting with a socio-environmentally just urbanism'. *Territorio*, vol 60, pp39–43.
Tornaghi, C. (2017) 'Urban agriculture in the food-disabling city: (Re) defining urban food justice, reimagining a politics of empowerment'. *Antipode*, vol 49, no 3, pp781–801.
Tornaghi, C. and Certomà, C. (eds) (2019) *Urban Gardening as Politics*. London: Routledge.
Tornaghi, C. and Dehaene, M. (2020) 'The prefigurative power of urban political agroecology: Rethinking the urbanisms of agroecological transitions for food system transformation'. *Journal of Agroecology and Sustainable Food Systems*, vol 44, no 5, pp594–610.
Tornaghi, C. and Halder, S. (2021, forthcoming) 'Reconfiguring the intersection between urban food movements and agrarian struggles: Building an urban political agroecology praxis'. In: B. Engels, K. Dietz, H. Akram-Lodhi and B. McKay (eds), *The Edward Elgar Handbook of Critical Agrarian Studies*. Cheltenham: Edward Elgar Publishing.
Vandervelde, E. (1903) *L'exode rural et le retour aux champs*. Paris: Felix Alcan.

Viljoen, A. (2005) *Continuous Productive Urban Landscapes: Designing Urban Agriculture for Sustainable Cities*. Oxford: Elsevier.

Viljoen, A. and Wiskerke, J. S. C. (eds) (2012) *Sustainable Food Planning: Evolving Theory and Practice*. Wageningen: Wageningen Academic Publishers.

Vitiello, D. and Brinkely, C. (2014) 'The hidden history of food system planning'. *Journal of Planning History*, vol 13, no 2, pp91–112.

Ward, C. (1999) 'A peopled landscape'. In: K. Warpole (ed.), *Richer Futures: Fashioning a New Politics*. London: Earthscan, pp83–98.

Wascher, D., Zasada, I. and Sali G. (2015) 'Tools for metropolitan food planning - a new view on the food security of cities'. In: M. Deakin, D. Diamantini and N. Borrelli (eds), *The Governance of City Food Systems*. Milan: Fondazione Giangiacomo Feltrinelli, pp68–97.

Wekerle, G. R. (2004) 'Food justice movements: Policy, planning, and networks'. *Journal of Planning Education and Research*, vol 23, no 4, pp378–386.

Wezel, A., Bellon, S., Doré, T., Francis, C., Vallod, D. and David, C. (2009) 'Agroecology as a science, a movement and a practice. A review'. *Agronomy for Sustainable Development*, vol 29, no 4, pp503–515.

Wiskerke, J. S. C. (2009) 'On places lost and places regained: Reflections on the alternative food geography and sustainable regional development'. *International Planning Studies*, vol 14, no 4, pp369–387.

Wiskerke, J. S. C. (2015) 'Urban food systems'. In: H. de Zeeuw and P. Drechsel (eds), *Cities and Agriculture. Developing Resilient Urban Food Systems*. London: Routledge, pp1–25.

Zasada, I., Schmutz, U., Wascher, D., Kneafsey, M., Corsi, S., Mazzocchi, C., Monaco, F., Boyce, P., Doernberg, A., Sali, G. and Piorr, A. (2019) 'Food beyond the city – Analysing foodsheds and self-sufficiency for different food system scenarios in European metropolitan regions'. *City, Culture and Society*, vol 16, pp25–35.

2

SHARING THE HARVEST

Transformative artful and activist methodologies for urban agroecology

Ella von der Haide, Anna Maria Orrù, Barbara Van Dyck, Debra Solomon, Mama D. Ujuaje, Deirdre Woods, Severin Halder and Robin Grey

Part 1: Introduction – Composing a speculative fabulation

Food is an entanglement of materiality and social practices, power relations, communities, histories and visions. It is a materiality that dwells on both the inside and the outside of the human body, in personal, public and urban metabolisms. With such a complex interconnectedness, how then can food, its politics and practices be researched in order to grasp the subject without losing this entanglement?

This chapter explores experimental methodologies and practices that hold the potential to shape the way we do research and communicate its results. This takes into account the complexity and heterogeneity of producing and digesting food, shaped by diverse histories, politics, interactions, relations, preferences, norms, beliefs and interacting organisms. If every phase of the food system shapes the interaction between human and more-than-human environments, what role can such research take in this interface, debate, dissemination and transformative potential?

We gather eight experimental practices that use modes of performance: dancing, cooking, singing, film-making, mapping, action-making, embodiment and growing. These modes encourage debate, activism and research around agroecology and food sovereignty. These practices help to access and circulate diverse knowledge bases that are unknown and non-valorised by normative and traditional forms of academic research. They allow individual scholar-activists to create political positionalities using performative techniques that promote reflexivity and invite people to participate in public and diverse dialogues outside of academia. As such, they allow access, validate and make terrains of discourse and change available to marginalised and unheard voices that wish to create different science, society and environmental relationships.

To enter into food knowledge dialogues, there are several conceptual feminisms that we take to help sift through the materiality and mattering to show what we build upon. The feminist approaches we use are string figure (SF), situated knowledge, fictocriticism and storytelling, and bodily practice, such as *makingthinking*. This chapter is revealed as an SF configuration to stay attentive to complexities, entanglements and different actors.[1] The feminist concept of SF – referred to variously as speculative fabulations, string figures, or scientific fabulations (Haraway, 2012) – inspired us to think about and play with food webs (replacing ideas of hierarchical food chains). Haraway (2012) proposes that by embracing metaphors of SFs to describe phenomena, situations and relationships in a scientific way, it is possible to include the ideas of different agents holding entangled strings before passing them on to other players. With SF, we think in other typologies and worlds wherein the borders of thinking, doing and making are shifted so that non-harmonious agencies become transparent and assembled in uncommon configurations that are 'multiform, at stake, unfinished, consequential' (Haraway, 2003, p.30).

In our first SF composition, we discuss how urban agroecologies and food sovereignty can be researched with support from experimental practices, such as growing, cooking, singing, storytelling, film-making and performance, and what advantages and challenges they create. SF belongs to a genre of fictocriticism and storytelling, which generate critical imagining and altered views of the urban social and ecological landscapes. Hélène Frichot et al. (2012, p.69) explain that fictocriticism, a feminist narrative style that weaves in fiction, theory and critical thinking, 'enables the expression of minor voices, construction of other possible worlds'. Furthermore, it gives the situated capacity to ethically cope with what confronts us' (*ibid*) and helps in understanding the complexities and entanglements of ecological manoeuvres in food-related urban matters (Orrù, 2017a). Fiction is fundamental because it is a way of sense-making an experience and can be retold. In this re-enactment, one ingredient necessary for approaching performative space is by telling an ecological narrative and creating urban fictions that make us think differently about others from an equal footing. Finding new fictions to live by is necessary if we are to develop approaches that incorporate justice into urban agroecologies. The work presented is also part of *makingthinkingdoing* practices in feminist artistic and practice-based research, which is inherent in feminist critical work. Natalie Loveless (2012, p.103) defines these practices in:

> What matters is our willingness to engage the multiple ways in which this ¨*making*¨ is a fundamentally situated, relational construct; one that entangles us in relations of debt in ways for which we can never account, despite always being willing to be accountable.

This means that the practice leans on the act of *making* and *thinking* whilst *doing*, used as a form of engagement and co-creation. The 'who' is making and co-making is equally important as to what is being made and how.

In the second part of this chapter, the SF thread is passed between eight participants and their projects presented in interview format and text submissions. These companions bring diverse modes of inquiry and patterns.[2] Robin Grey (UK) has created a show that covers a thousand years of English land rights history in folk songs, poetry and story, with an accompanying open-source performers kit. Ella von der Haide (Germany) makes films on and with community gardeners and their entanglements with social movements working on ecological justice, post-colonial questions, feminisms and queer-feminist politics. Severin Halder (Germany) and *kollektiv orangotango* organise collective mapping processes in which participants learn about powers that are embedded in maps and learn how to influence and speak about them. Anna Maria Orrù (Sweden/Italy) creates interventions using *fictocriticism* and *makingthinkingdoing* practices, culminating in a Butoh performance exploring the interplay between food and body(ies) in regard to transformative pedagogical methods in sustainable urban-making. Debra Solomon (the Netherlands) and the *Urbaniahoeve* collective plant park-like food forests and create fertile (top)soils as an interactive, interspecies communal space and new form of public space.[3] Mama D. Ujuaje (UK) enacts the interactive and multi-sensory '*Food Journey*' that guides blindfolded participants through imperialised food and food/body common histories, disrupting both to make way for embodying public agency. Barbara Van Dyck (Belgium) develops and performs a 'lecture in motion' or 'lecture-performance' about agricultural research as an issue of public concern. Deirdre Woods (UK) cooks with groups as a gateway to discuss food, food systems, democracy and colonial politics. The third part ends with a reflection on the eight practices and paradigm shifts for such models for knowledge production.

The practices presented come in different configurations; some use art and creativity directly embedded into research (*Grey, Haide, Halder, Ujuaje*), some create art as a form of research (*Orrù, Solomon, Van Dyck*) and some use artistic expressions to communicate and question research and to politicise scientific underpinnings (*Haide, Orrù, Van Dyck*). A few participants come from academia where they began using artistic methodologies (*Orrù, Van Dyck, Halder*), while others emerge from artistic and community backgrounds, and started collaborating with universities to assemble alternative approaches to knowledge production (*Grey, Haide, Solomon, Ujuaje, Woods*).

Creating experimental methodologies

We use the term 'experimental methodologies' to describe our practices as they veer away from traditional approaches to urban agroecologies. Each of these works utilises artistic and creative approaches to highlight alternative forms for participation, communication and dissemination *in/with* urban food systems. The *experimental* is about how we bring together art, research, activism, collective learning, sciences and society into a collective experiment (Latour, 1998). It also refers to the way in which we start to create new dialogues between humans, more-than-humans and the environment. All participants' projects are modes of activism with a political and emancipatory agenda. Some work with collectives on food sovereignty and agroecology to transform cities (*Haide Halder, Solomon*), while others intersect research processes to demonstrate alternative forms of knowledge production (*Grey, Orrù, Ujuaje, Van Dyck, Woods*). Hence, they demonstrate the need for a diversified *making* and *understanding* of knowledge.

The experimental methodologies presented here share some similarities with *participatory action research* (PAR) that uses creative and hands-on methods to make research tangible, meaningful and accountable (Ayala and Zaal, 2016; Fox, 2015). In some projects, PAR helps to designate a research intention and illuminate a specific problem (*Halder, Solomon*). Others are interested in raising questions or awareness, and changing the way in which knowledge is created and circulated (*Haide, Orrù, Solomon, Van Dyck*). A third intent is working on general social structures, beliefs, knowledge and histories (*Grey, Ujuaje, Woods*). Furthermore, all are involved in the actual *matter* of what we discuss and reflect upon, whether it be growing, gardening, composting, or cooking. The approaches are not only concerned with defining challenges that are valid for social movements, ecological and social justice, but also to generate learning and research data using emancipatory methodologies and respecting different ways of knowledge-making.

All examples enter into dialogue with people who do not necessarily share professional, disciplinary, or cultural backgrounds with scientists in order to make the knowledge available to different audiences. This chapter points towards not only bringing together different practices of collective experiments and activism, but also showcasing alternative modes for the way *knowledge* is produced and understood in academia, at the same time, acknowledging that knowledge produced outside academia is equally valid. From the perspective of knowledge production, two forms of research are extending to the presented projects and their methods: *artistic research* and *participatory action research. Artistic research* connects to our work wherein art or an artistic process is used as a way to think with, connect and make relationships, and to communicate critical ideas to a wider audience by involving them in urban food systems.[4] It can also be used as a methodology for norm critique exploring varying forms of practices, encounters and relations. Catharina Dyrssen (2015, pp.23–24) states that

> [a]rtistic research often takes as its starting point the scope for art to shed light on value-related and communicative questions concerning creation of meaning and quality. The research methods tend to be action-oriented, performative and interactive with artefacts and the surroundings.

Both artistic practices and research can allow for critical visions, emotions and ethics to play a larger role in finding site-specific clarification, including voices that are marginalised. Research into food, food cultures, politics and practices must be aware of other forms of knowledge-making. These knowledge practices embrace a non-hegemonic view of nature (*naturecultures*) by merging together ecology and art, especially since food is a materiality in encounters with people (human and more-than-human), land, dirt, creativity and plants.[5] Artistic research is risky, but this consequence is part of the doubt, unknowing and discomfort inherent in experimental methodologies. It allows us to have direct experiences and generate empathy, more so than any written report/statements. At the same time, it might also help us transform food experiences to gain hope and initiate action. In order to go beyond only traditional modes of approaching urban agroecology discourses, we need to place ourselves in situations of unknowing and discovery. In staging such profound modes for approaching and transforming agroecological urban practice, Erika Fischer-Lichte (2009) calls for creating a *liminal* experience – to know beyond what we already know – wherein the process of unknowing oneself, the knowing of oneself happens again, and one is transformed (Orrù, 2017a).

Potentials and challenges of experimental methodologies in urban agroecology

The paradigmatic changes that urban agroecology necessitates in terms of science-society relations, humans and human settlements positioning within food systems are too important, too entangled, too political to be left to the reified depoliticised practices of (neoliberal) scientific practice. Therefore, embodied and situated approaches to an urban nature have meaningful consequences that also link to ethical behaviour and justice inherent in place attachment, involvement and care. In such an orientation, we situate the body as an anchor of inquiry and relation-making from which to approach environmental challenges. Orientation is about relationality, because which way we orientate determines who or what we decide to relate with and to (Ahmed, 2006). Therefore, the difference between 'knowing of' and 'knowing how' rests in mobilising the creativity of the body into playful encounters, which bring this situated 'know-how' to the surface through situated knowledge (Orrù, 2017a).

A main knowledge grounding present in all the showcased projects is Haraway's (1988, p.585) concept of *situated knowledge* that reasons for a positioned and embodied knowledge. She calls for a 'doctrine and practice of objectivity that privileges contestation, deconstruction, passionate construction, webbed connections, and hope for transformation of systems of knowledge and ways of seeing' (Haraway, 1988). This means a turning away from dichotomies of body and space, human and non-human, urban and rural, and nature and city, to see their inherent complexity of webbed accounts, multiple knowledge, heteroglossia and ethnophilosophies (Orrù, 2017a, p.100). Furthermore, Haraway argues for knowledge that is entangled and emerging, 'for the view from a body, always a complex, contradictory, structuring, and structured body, versus the view from above, from nowhere, from simplicity' (Haraway, 1988, p.589). Moving towards a transformative food-enabling city includes a change in values and norms that welcome other forms of knowledge-making. Art allows room for these values to be discussed, questioned and critiqued openly in society. Poethical practices encourage ethos both as a poetic and political embarking, and reconfigure ethics into a mode of encounter and relationality (Orrù, 2017a).[6] With creative processes we begin to reimagine worlds, probe alternative urban existences, and catalyse change (Haide, 2003–2018). The transformative examples we present shape their internal organisation and practices in a visionary way so that they are already 'laboratories' for multifaceted societies. This is one reason for staging experimental methodologies outside academic contexts in order to gain freedom in the processes and to work around norms present in academic structures.

In each case, situated knowledge is viewed as entangled 'action possibilities' stemming from embodied practices. Situated knowledge is a broad umbrella under which other forms of knowledge are also present and particular to each case. In the eight contributions that form this chapter, we identify the following knowledge and its emergence:

- knowledge so common, it is uncommodifiable, devalued and overlooked. This knowledge is about reproduction in households, e.g., everyday cooking or gardening (*Ujuaje, Solomon, Woods*);
- knowledge that reclaims and revalues real cooking from scratch, and reestablishes the connections it makes with others and where the food comes from and who produced it (*Ujuaje, Woods*);
- knowledge suppressed or forgotten, but connected to emancipatory struggles over time, e.g., land rights and colonialism (*Grey, Ujuaje*);
- untapped embodied knowledge that is often devalued, tacit, or inaccessible through normative means but determines how one orientates in a *poethical* – poetic, political and ethical – stance (*Orrù, Ujuaje*);
- knowledge discussed in rational sciences, such as alternative ecologies and cosmologies, that exist in gardening (*Haide, Solomon*);
- knowledge that marginalised and oppressed people possess, e.g., queer approach to ecologies (*Haide, Ujuaje*);
- knowledge from lived diverse *naturecultures* (*Haide, Orrù, Ujuaje, Woods*);
- knowledge that is uncomfortable to talk about, such coloniality that is embedded in food systems (*Ujuaje, Woods*);
- inaccessible knowledge because of the sense of helplessness, such as awareness of climatic changes (*Orrù, Haide*);
- local knowledge that is devalued because it is not considered expert (*Haide, Halder, Van Dyck*);
- visionary knowledge that builds stories about alternative futures (*Haide, Orrù*);
- knowledge relating to emancipatory struggles, organised commons, independences and what a world with less injustice could be (*Haide, Halder, Ujuaje, Van Dyck, Woods*).

In the process of creating works more independently lies the challenge of having to sell or commodify them in other ways. Therefore, the dilemma is to present relevant work that is non-consumable but understandable and accessible, while bringing up inconvenient questions needing discussion. Daring to create discomfort and confusion signifies that learning and transformation is taking place. In addition, as much as these experimental methodologies can empower, they also open up new forms of marketisation and expropriation. Experimental methodologies might create a new pressure on the single researcher to be both a performer and an action researcher in order to reach out to all these wider-reaching actors. To get funding, researchers often have to develop marketable and solution-orientated questions, processes and results, and presentations.[7] The challenge is how to enhance the value of these forms of knowledge while building in mechanisms that complicate their misuse, theft, or commodification. One answer may be to transform the new knowledge directly into a *commons* and make it publicly available (*Grey, Haide, Halder, Solomon*), edible (*Ujuaje, Woods*) and performable (*Orrù, Van Dyck*). Such public orientation provides increased discussion by repeatedly positioning it in political spheres and giving veto to neoliberal exploitation.

Part 2: Weaving an SF thread composition

In spirit of our SF game, we offer a variety of artistic contributions that are presented in interviews (*Halder, Solomon, Van Dyck, Woods*) and other textual formats (*Grey, Haide, Orrù, Ujuaje,*). At this time, we pass the SF thread around between all eight participants.

Thread 1 – Robin Grey

Three Acres and a Cow – a thousand years of English radical history through folk songs

Three Acres and a Cow is a site-specific, improvised performance based around a modular open-source creative-commons framework.[8] The show covers a thousand years of English social and radical history through folk songs, poetry and stories. The multiple histories of past resistance and activism are largely undocumented in the literature of the time, often expressed simply as a hanging here and an uprising there, yet in the music and stories of the people they take on a different life. It raises awareness of how present land use and ownership patterns in the UK and beyond are the result of policies and decisions that have little basis in social justice or in consideration of the common good. The project has unearthed primary source material, which has never been published or performed, inspired new projects and collaborations, and reinvigorated existing work.

The show is a six-year-old creative action research project, combining what I learned from working in community food growing projects about politics, land rights and alternative models of training/organisation with my passion for creative-commons and open-source intellectual property ideas from the computing world. We encourage audiences to join in with the songs and everyone gets a lyrics booklet, which they are free to take home at the end. Through this, we support people to be active participants in the show rather than passive consumers. We crowdsource information, songs and stories from audience members after shows. An online wiki 'performers' kit' allows anyone to put on a show without needing trained practitioners. We keep ticket prices as low as we can, and take performances into spaces that are already part of people's lives, weakening the traditional arts venue performer/audience member divide. The project has led to the creation of spaces where diverse communities of interest have had critical discussions on important issues raised by the project, including land rights, housing rights, food sovereignty, public space, planning, farming subsidies, climate change and the environment. The show is used by community food growing projects, housing campaigners, academics, faith-based groups and social justice activists as an outreach tool for explaining the context and importance of their work and convictions to local communities, and empowering people with often forgotten historic narratives.

The project supports local interest groups so they can co-host shows and encourage the presence of stalls at the performance, allowing space for networking. Each show also creates space for further discussion with stalls and generous space before and after performances, either on the occasion of a performance or with a follow-up workshop, to go deeper into the subject matter. Evenings often start with a simple shared meal, increasing the amount of time and space for people to engage, with systemic ideas around land rights and food systems. We embed feminism in our practice by including as many stories of empowered and brilliant women across the ages as we can find, and making sure that there is gender diversity among the shows' core performers.

The image (Figure 2.1) is an example of one 'song' reaching out. We believe that increasing nationwide land literacy will lead to a mass movement around land rights in the UK and to a more just distribution and control of land.

Thread 2 – Ella von der Haide

'Another world is plantable!' – Films on community gardening around the world

Film-making is an art, and alternative media is a form of activism. I work *in*, *with*, and *about* community gardens in different ways. 'Another world is plantable!' is my series of six films on community gardens and their connections with social and ecological justice movements in South Africa, Argentina, North America and Germany. I have produced films in this genre since 2003 (Haide, 2003–2018, forthcoming), including two films on the resistance to the European seed legislation.[9] In my documentaries, urban gardeners and farmers explain how their gardens feed both them and their communities on the physical, social and psychological levels. Simultaneously, they serve as instruments in their anti-neoliberal, feminist, food sovereignty and social and ecological

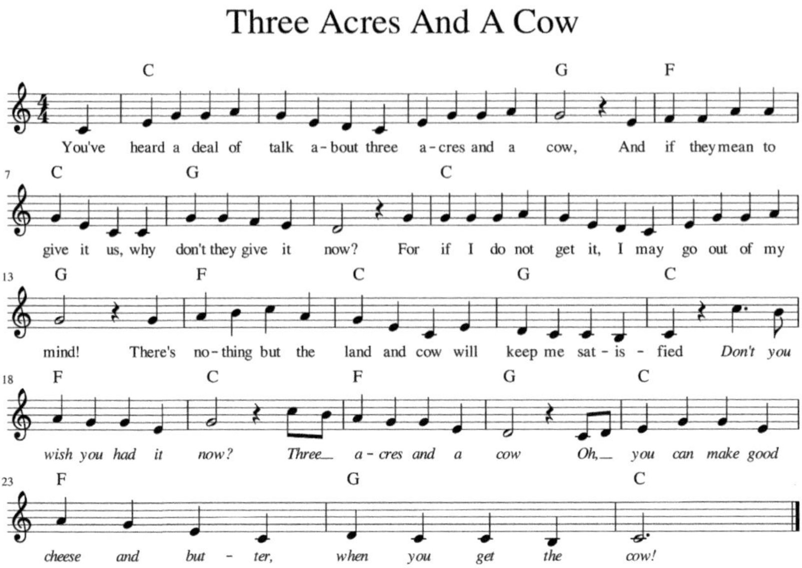

There's a certain class in England that is holding fortune great
Yet they give us all a starving wage to work on their estate
The land's been stolen from the poor and those that hold it now
They do not want to give us all three acres and a cow

D'y' think they'll ever want to give three acres and a cow
When they can get a man to take low wage to drive the plough
To live a man he has to work from daylight until dark
So the lord can have both bulls and cattle grazing in his park

But now there is a pretty go in all the country though
The workers they want to know what the government will do
And what we have been looking for, I wish they'd give us now
We're sure to live if they only give three acres and a cow

If all the land in England was divided up quite fair
There would be work for every one to earn an honest share
Well some have thousand acre farms which they have got somehow
But I'll be satisfied to get three acres and a cow

This song from the late 1800s would have been sung by agricultural labourers who'd only just won the right to unionise and had directly experienced or at least had strong memories of the Enclosures. The Irish had won some significant victories for land reform in recent years and many people in England, Scotland and Wales were inspired to attempt similar feats on their island.

'Three Acres And A Cow' was a slogan of the land reform movement which was led by a number of Liberal MPs. Instead of getting three acres, the labourers were fobbed off with the allotment acts which gave them 1/12 of an acre and no cow.

FIGURE 2.1 *Three Acres and a Cow* song and music. *Source:* Robin Grey (used with permission).

justice struggles. My films aim to raise awareness about existing alternative projects, to create visions for transformation to oppose the neoliberal '*there-is-no-alternative*' logic, and to motivate people to engage in community gardens. They feature activism and are used by activists to mobilise and raise consciousness.

The films are also a part of my academic work at the Department of Urban Planning at the University of Kassel, where I use audiovisual, artistic research and communications methodologies. In my scientific work, I am involved in the development of urban policies to support and expand urban gardening in Germany. At the university, using my films, I teach future urban planners and architects to create awareness within their professions for participation, grass roots projects and ecological ideas of urban agriculture. It is my experience, the use of film is helpful in all these areas to create a better understanding of the phenomena, local situations and to create empathy. Audiovisual methodologies have the ability to answer to other types of learning. They create a sense of place and transfer local knowledge that would be difficult to convey through only written documents.

Another central goal of my work is to create critical political awareness within the urban gardening scene itself. Through my film-making, gardeners can enhance their media knowledge and gain sovereignty over the image production around the gardens. Since community gardens are subjected to appropriation, through green washing and image campaigns used by companies or the city management (Halder, 2018), the creation of self-determined images and stories becomes a means of resistance. I interview people who would not usually be heard, allowing for these marginalised voices and views to enter the poetical debates. My hope is to empower garden activists and represent them respectfully.

The documentaries convey knowledge about potential alliances between gardeners all around the world. Inspired by feminist writers Donna Haraway (2016) and Anna Tsing (2017), they cultivate a way of attentiveness and an 'art of noticing' and to stay with complexities and troubles. With my approach, I show environment-society relationships that are not based on heteronormativity and patriarchy and queer ecologies (Haide, forthcoming). My work has been formed by Haraway's work on cyborgs and naturecultures, Chthulucene and composting, Jane Bennett's (2010) 'new material feminism' and Mortimer-Sandilands and Erickson's (2010) queer-feminist ecologies. These alternative ecologies and non-hegemonic relationships with the environment are based on and create different kinds of knowledge, which are underrepresented, in scientific literature. For example, the idea that a non-human world looks back (Haraway, 2003), that humans can tune in to hear the messages of plants, or that gardeners can coevolve with the microbiom to create soil are not accepted in Western rational views but can be found amongst gardeners. Some of these ideas are traditionally passed on and some have developed recently. These alternative relationships with ecologies hold the potential to open up new ways to structure our paradigms about *nature-cultures* on which we base academic research, the organisation of settlements and food production, consumption and reproduction.

A challenge I face is that it is always a compromise to edit the multi-layered interviews so they can be taken in by the audience in the limited time frame of a film, while staying close to the complexities. When my work is described in newspapers or on the radio, the message is very often reduced even further.

Another difficulty is that movies do not invite reflection from the audience. Films create emotions – sometimes people cry when they see my films; this makes them good tools in creating empathy and for learning. At the same time, however, these are strong experiences that do not catalyse people to question things. For my film screenings, I try to always follow them with talks – question and answer sessions.

Thread 3 – Severin Halder

Collective mapping

Please describe what you are doing/creating?
As part of the *kollektiv orangotango,* I am involved in geographical activism. We mainly organise collective mapping workshops within the contexts of urban agriculture and 'right to the city'. *Kollektiv orangotango* seeks to support processes and oppositional actors who instigate social change by prefiguring social alternatives. We conduct educational work as well as concrete political and artistic interventions with the aim of collectively learning how to read space and how to initiate emancipatory processes from below. In our mapping workshops, we are generating – in a ludic, creative and productive way – maps on different scales: from how to organise resources in a single community garden to finding allies and projects of an alternative food system in a region. Through the means of drawing a map, a horizontal dialogue is initiated and counter knowledge is socialised and made available in a creative and visual way. The results of this mapping can vary from collective scribble maps to fancy critical (online) maps of all the community gardens in a region. But even more important is the education process, which we describe as a collective process of geographical alphabetisation in spaces of everyday action. Because drawing a map together involves a way of creating collective accounts on what is common to us all and this builds a platform rendering visible certain contacts and consensus without reducing diversity, for this is also depicted (Ares and Risler, 2016, p.10). At the same time, the participants appropriate the knowledge of mapping and reading maps critically; this is crucial to understand and influence, for example, urban planning.

The *kollektiv orangotango* is a space for action and reflection, activism and art, and geography and friendship. We are working on critical education and supporting self-organised structures within the context of everyday resistance, emancipatory struggles and social movements. Inspired by activist geographies, counter cartographies from Latin America and the 'pedagogy of autonomy' (Freire, 1970), we want to create solidary relationships and horizontal knowledge exchange.

In what way is it art, research and activism?
I am an activist and critical geographer who holds a PhD (Halder, 2018); that is why I seem to be more academic, but my work always includes creativity,

whether it is publishing, mapping or growing. Collective mapping is a method that includes creativity in the form of drawing and activism in the form of creating counter knowledge and territorial appropriation. It has also proven to be a good tool to use in activist research to talk about a situation, show power relationships, common uses and visualise alternatives.

Why did you choose this artistic methodology to work in the field of transforming food systems?

My work is inspired by the everyday resistance of urban peasants whom I met in Rio de Janeiro, Bogotá, Medellín and Cape Town, and friends and comrades like *Iconoclasistas* in Argentina and *Nova Cartografia Social da Amazonia* in Brazil. Another inspiration comes from all different experiences with creative forms of resistance, street parades, street art and interventions in public space. I started with Guerilla Gardening, and I was astonished and amused by the huge impact that you can have by throwing some supposed to be illegal seeds into public green space. I am convinced that you have to choose any kind of emotional, funny, artistic, or haptic language to reach out to the people you want to talk to.

What kind of transformative effect are you hoping to achieve?

I hope to make the world a tiny bit better by being part of a collective solidary, ecological and emancipatory transformation process. I see myself, therefore, on the one hand creating concrete tools and places, and on the other hand shedding light on the contradictory potential of activism and action research, e.g., in urban gardens or collective mapping processes.

Within the context of urban gardens, I am involved in a combination of networking, education and action research by creating spaces of encounter, critical self-reflection and practical tools such as handbooks, maps or seed exchange to create more political debates.

What are your experiences in working with artistic methodology?

A lot of fun and wonderful possibilities to combine different spheres of knowledge and creativity.

How does it create new forms of public dialogue and debate?

Maps are tools of communication and so they are powerful means to influence public debates.

Do you feel you are becoming a performer and that you make resistance and transformation consumable through your work?

Our approach within *orangotango* is to create dialogues by being engaged in processes of popular education that brings us to make theory, resistance and transformation accessible.

Thread 4 – Anna Maria Orrù

Corporeal encounters with farmscapes – embodied methods for ecological urban-making[10]

The intervention *Organoleptic Interfaces* brings elements of space and urban agroecology into an embodied and performative position. It has two modes of inquiry: a *makingthinkingdoing* modelling workshop, and a performance that uses choreography as a means for harnessing situated knowledge. The aim is to set up an encounter that activates alternative attentiveness, profound reflectivity and stirs imaginations. These approaches belong to the critical feminist spatial practice that allows ways of confronting urban food issues by setting up alternative embodiments with nature. It does this by developing methods that promote deeper understandings of the knowledge we use.

To incorporate speculative fabulations, I ask participants to physically model text on natural farming and employ Butoh choreography as a form of knowledge transfer.[11] The first mode, *Paperscapes*, includes a knowledge-assembling workshop delving into Masanobu Fukuoka's book on natural farming (Fukuoka, 1978). Paperscapes is an active shaping of his writings into a trace-paper sculpture stage, which is used afterwards for the performance. The text was chosen based on Fukuoka's perspectives on nothingness, his farming method, natural approach, seasonality/time, the body's role and life world(s). Fukuoka believed that non-action was the best way to produce fertile, healthy soil, and is in keeping with the laws of nature by generating a *do-nothing farming* that relied on nature to do the work. His guiding principles on natural farming include no tillage, no fertiliser, no pesticides, no weeding and no pruning, and growing diverse produce year-round, therefore complying with and replenishing nature's richness (Fukuoka, 1978). At this stage, the students self-organised into three groups representing farmscape boundaries: air, ground and edge.

Our body holds a situated knowledge that becomes activated through active interaction. For example, the knowledge of touching soil, and of the medium's air, wind and weather, can become accessed if the body remains attentive to how it interprets and reacts. Once every participant recognises these properties, the added imaginary leaps to grasp natural farming is not far (Orrù, 2017a). Tim Ingold interprets this tacit knowledge as the 'surface of our life-world', that such knowing 'carries the intent of creating a hard boundary between what lies below and above, and metaphorically between the material and the mental' (Ingold, cited in Marchand, 2011, pp.15–16). In this mode, the knowledge emerges through the *makingthinkingdoing* process of sculpting text into an abstract garden.

Next, Butoh-dancer Frauke performs in the papercrafted farmscape. Unrehearsed, she enters the space for the first time. The intention is to bring her body into dialogue with the various constructed metaphors and materialities of the space. She uses choreography as a form of communication where space prompts body and body animates space. These are interfaces in between real and imagined worlds – a biological fiction in a paper world that calls for deepened visceral practice and ecological understanding through performing *natureculturess*.[12] Frauke uses Butoh as her spatial language with techniques that include growth, spiralling, movement from ground to air and decay.[13] These *bodily* properties activate voids around Frauke's body allowing her to discover the stage through her *body* in motion and relation with the space. Her senses become measuring devices and each world created is conditioned by her reinterpretation of the constructed 'garden'. Elin Diamond (1996, p.2) writes that such performance skills are an important part of transformative knowledge processes 'to "embody", "configure", "inscribe", "signify", assert the possibility of materialising something that exceeds our knowledge, that alters the shape of sites and imagines other as yet unsuspected modes of being'. In imagining something that is not there by materialising its presence through movement and engagement, the space changes, as does the body.

Outcomes include reformulated aesthetic pedagogic methods for curating corporeal and embodied poethics into urban-making. Such transformative learning stemming from *makingthinkingdoing* practices allows the transfer of knowledge through a bodily practice, whereas traditional methods tend to create a separation between them. The intention was to explore ways in which a trained perception of materiality could emerge to better understand what happens when a body is engaged with the ecological rhetoric used to shift urban lifestyles. Using SF as a form of metaphoric narrative to transfer knowledge as a performativity helps grasp the complexity of environmental challenges, suggest behavioural options and catalyse action. It becomes a political act that breaks with traditional education in architecture, by bringing body and performance to the fore (Orrù, 2017a). Deborah Hauptmann (2006, p.10) understands the use of the body not only as formative but also transformative, both practically and theoretically. The way in which the world is conceived creates a mode of behaviour where reasoning, desires and knowledge belong to the realm of corporeal affects. If what the body does is connected to reason and what we know, and *vice versa*, then approaching matters of environmental behaviour must include the body (Carolan, 2009, Orrù, 2017a). Urban-making needs to develop a profound corporeal depth of methodological expertise that includes the body and acknowledges corporeal presences as integral to making, understanding and composing space.

In summary, finding modes of pedagogy where students model environmental knowledge can become transformative processes that enable an experience-based comprehension of literature on space and food. The reading of Fukuoka's text was enhanced by *thinkingmakingdoing* practices intrinsic in the sculpted paperscape and enactment. It was important to remain reflexive to the task at all times so that the actual act does not simply become a seductive aesthetic and designed experience. This requires the workshop to be critically attentive to how the assignment is conceived and its implications for learning. Hence, under a rigorous pedagogic, these forms of critical feminist spatial practices could have huge shifts in the way students learn about practices in urban agroecology.

FIGURE 2.2 Film reel images for Organoleptic Interfaces (Orrù, 2017b). *Source:* Anna Maria Orrù (used with permission).

Thread 5 – Debra Solomon

Urbaniahoeve technosols – Multi-species urbanism, public space topsoil production and portraiture

Please describe briefly what you are doing/creating?
In 2010, as an artist/researcher, I founded *Urbaniahoeve* (in Dutch, 'the city as farmyard'), a collective collaboration with locals to plant park-like food forests in urban public space (Solomon, 2012; Solomon and van den Berg, 2014). To create urban food forest ecosystems, we nurture the soil organism, establishing rich topsoil from urban waste streams. The ecosystems and resulting *'technosols'* are a platform used to facilitate new socio-ecological relationships between the human stewards and more-than-humans (Solomon and Nevejan, 2019).[14]

For the art series *Soil Portraits,* I innovated a soil chromatogram magnification (16×), using chromatography techniques that communicate properties of each soil.[15] Applied widely in DNA research, chromatography's analytical value is contested within the culture of soil science.

In what way is it art, research and activism?

Technosol production as artwork references the land art tradition, especially Smithson's entropy works. The proposal for public space food forest ecosystems, including the aforementioned *technosols*, is activism, positing new formats of interactive, multi-species public space production and collective steps towards urban food sovereignty. The research locations (art, social science and soil science) are also case studies for my PhD on multi-species urbanism.

Why did you choose this artistic methodology to work in the field of transforming food systems?

Soil Portraits (Figure 2.3) applies chromatography to amplify soil fertility data within a visual medium, allowing it to travel to more diverse, non-traditional audiences. The results of the scientific data within these *chromatograms* provide information beyond quantitative analysis. Within this oeuvre, the contested legitimacy of chromatography as a soil analysis methodology takes the role of provocation.

What kind of transformative effect are you hoping to achieve?

The soil organism as a whole, particularly these self-made *technosols*, features prominently in my work as a utopian tool, grounds for Commons, as a biotope provocation. Technically capable of altering the course of our climate crisis, soil organism care measures our ability to nurture and repair ecosystems.

Democratically developed and maintained public space that provides nutrition sovereignty, ecosystem-based climate adaptation and focusing on the temporalities and rights of urban ecosystems will become a new form of public space altogether.[16]

What are the advantages and challenges of working with artistic methodology?

An advantage to working with artistic methodologies is the ability to tread outside disciplinary and political categories, which in turn creates the challenge of maintaining comprehensibility.

Are you accessing new forms of knowledge production? How?

The *Radical Observation,*[17] methodology develops observation and natural dynamics awareness in intervention landscapes. New perceptions towards socio-ecological relationships guide our behaviours, thus forging new forms of governance and policy.

Do you have thoughts about other artistic methodologies within food sovereignty? What other examples do you know of or want to see?

Sue Spaid's *Ecovention Europe* exhibition (Spaid, 2017) and Alexandra Toland's *Field to Palette*, (Toland et al., 2018), to which I contributed.

FIGURE 2.3 Soil Portrait #26 (Debra Solomon, 2016, 580 × 580 mm) is a chromatogram sampled from topsoil produced in situ by Solomon in close collaboration with the non-human soil-building community of the Urbaniahoeve Food Forest Ecosystem in Amsterdam Noord. The artwork depicts a young, well-aerated soil, comprised of abundant, partially decomposed organic material. It illustrates that organic material is accessible to many species of soil life, even when not completely decomposed. *Source:* Debra Solomon (used with permission).

What aspects of feminism are important for your work?
Topsoil production and ecosystem stewardship is about observation and care (Puig de la Bellacasa, 2015), embodied *use value*; the terrain and bane of women.

Thread 6 – Mama D. Ujuaje[18]

The Food Journey – A sensory reflection on dimensions of food

We wait in anticipation. There is mounting excitement at the waft of aromas and peek of the sights within. An antechamber of expectation, we exchange our ideas with each other until the journey guides invite us in, one at a time, a gateway, now ritual embarkation point.

Seated, we don the darkness about our eyes and there is a shushing and smudging of footsteps. Our bags tucked in and the tartness of lemon on our lips, we await further instructions.

This is the beginning of the Food Journey: an immersion in a 'hidden histories' narrative of what we consume and why we consume it and where, folded between the lesser told tale of the commodification of humans and how this led to the global commons of foods we know today.

Part of the inspiration for the journey arose out of the frustration with the linear, unidirectional, single narrative around what is referred to as 'the food system'. It is not possible that this sensual, evocative sustenance that is global food can meet our nutritive, digestive, societal and emotive needs if fed to us in small dry morsels, without taste, aroma, or memorable strains of music.

The so sumptuous nature of the world of cuisines and rare recipes is such that it demands storytelling in the old traditions, accompanied by multi-sensory stimuli. We are transported each step of the way in the manner of the food itself: textured, emotive, aromatic. We create, with a *griot's* finesse, entry into multiple dimensions that themselves problematise the dry textual accounts of origins, sources, biologies and economics found on dusty university shelves. We literally re-search the connections, ambivalences, contradictions and the veering off the beaten path of food stories. We encourage participation because we don't trust that 'hunters can be trustworthy historians of the hunted'.[19]

What is the 'soundtrack of enslavement'? What colours and tastes should accompany the heat and fetid rawness of salt-encrusted souls buried deep in the ship's hold? What is the exact texture between tongue and palate that connotes travel from east to West Indies and the stirring of *cassareep* into the *pepper-pot*? First-hand accounts are buried as deep as Taino *barbacoas*, deep in the dusty, windswept soils of colonial histories – but the bones are beginning to show and the truth will, and must emerge – to the sighing strains of lost plantations and fortunes made from abolitions.

The Food Journey requires reflection. Not once but severally. Sharp, poignant, a sudden flavour in the mouth, an acrid burning smell followed by a spewing of seawater over the brain.

Is it not simply a staging point of knowledge seeking?

Hush.

Silence is broken by the ecstasy of rapid fire whispers and exclamations upon seeing what one has accumulated in one's lap? Where are the answers to the questions still stuck in the pre-throat-place of the one who still regurgitates the first question, asking what were the textures of his father's breakfast as a toddler.

What constitutes refuge?

The journey for some is a one-time escape to the restaurant of recollections. Where is the exit? The participants can also cook. If wished, they can re-purpose everything the journey has unfolded for them, and there is no insistence on how to particularise the response. Artists might become activists and activists artists, each seeking to redress the balance of their understanding and to trust the world again.

Such spaces are as rare as white maize in an aid portfolio. To enable them to be as useful as possible, we encourage the largesse of a plantation owner who is impressing the incoming new governor with the new foods of the colony.

Women were the geniuses who worked below the great house horizons, often with brand(ed) new materials/chattel, new species, imported, rare or grown for the purpose of staving off hunger. Kitchens were the anchors of staying power, sites of everyday invention, resilience and survival. Places from which the master might be served his just desserts and justice might be secreted out in hidden bags of surplus ingredients. Here, the very first food waste was used to feed the deserving poor. Arbiters of nutrition yet close at hand for breeding, privy to the great house secrets and custodians of cuisines.

Now, disembark. Step down onto a steady shore. Regurgitate the best bits, if you must. You may have been seasick, bring up what you have learned, make it a regular practice to revisit what is hidden. Empathise. Revolt! Make the journey a hoe to unearth all that you have learned to date!

Thread 7 – Barbara Van Dyck

Reclaim research – A lecture in motion or lecture-performance

Please describe briefly what you are doing/creating?
I developed and performed an 80-minute-long 'lecture in motion', or 'lecture-performance', about agricultural research as an issue of public concern.[20] Drawing on my experiences with citizen's collectives that actively claim their part in GMO-research experiments in Belgium,[21] and as a researcher employed at the university, the lecture-performance aims to create an experience that demonstrates, as Sheila Jasanoff and others have developed extensively (Jasanoff, 2004), that knowledge are both the product of and constitutive of social and ecological life.

In what way is it art, research and activism?
The lecture-performance approach does use conventions both from lectures and from theatre and performance. Public storytelling – including the spoken word as well as movement, sound, music, picture, video and props – is used to explore a political question in depth and weave strings of 'cold' and 'warm' knowledge intimately together. People from the Belgian collective of lecture-performers have their roots in popular education and present in different settings, including schools, cultural centres and civil society spaces (*Conferences Gesticulees*). Performances are usually followed by a workshop or debate.

Why did you choose this artistic methodology to work in the field of transforming food systems?
The format resonates very well with the radical critique on science communication and instrumental public engagement approaches that I seek to develop.

I take the question of knowledge co-production very seriously. Agricultural research is 'neither a simple reflection of the truth nor an epiphenomenon of social and political interests' (Jasanoff, 2004). The lecture-performance for me is a means to propagate the co-production framework as an analytical lens, as well as an act of creating the conditions under which it becomes possible to open up knowledge production processes to include hitherto unheard voices. For example, how to create the means for people to engage with the agricultural sciences? This may sound obvious, but it is not part of most science communication in which the idea is to hand over the truth to people. Neither do participatory research approaches, in which people are invited to participate in data construction of very specific questions that researchers have defined for them, create the conditions for epistemological and ontological changes with regard to who is in the knowing.

In addition, by broadening seemingly narrow technical questions into deeply political ones,[22] the very attitude of attributing 'all knowledge that counts' to the technical experts is questioned.

What kind of transformative effect are you hoping to achieve?
I hope to inspire people outside universities to engage with agri-food knowledge systems, as a crucial element in the larger struggle for social justice and ecological health, and to ask inconvenient questions that disrupt and transform existing science-society relationships.

How does it create new forms of public dialogue and debate?
The hybrid format of a lecture in motion allows the sharing of themes outside their usual circles and networks. Talking about science in a theatre in the framework of a festival on 'feeding the city' in Liege (Belgium), for example. The very animated panel discussion that followed consisted of farmers and civil society actors discussing action research.

Do you feel you are becoming a performer and that you make resistance and transformation consumable through your work?
The academic arena is already very much one of appearance and performance. Performing a lecture in motion is a way of playing with that, overdoing it. Clowns are an important source of inspiration here.

But the question is crucial and necessitates permanent self-reflexivity. With the Belgian collective of lecture-performers, we have had a lot of discussions about the focus on individual stories. For example, starting from one's life histories is empowering and creates transforming experiences, but how does it contribute to a culture of individualistic performance and 'Stars'?

What aspects of feminism are important for your work?
I owe a lot to feminism in my work, situated knowledges and (eco-)feminist science critiques in particular. By interweaving personal stories, play, humour and political ecology, the lecture-performance seeks to disrupt the colonial patriarchal and capitalist legacy of knowledge monopolies.

Thread 8 – Deirdre Woods

Cooking up revolution – recipes for food democracy

Please describe briefly what you are doing/creating?
I host interactive cooking classes using a combination of vegetables, spices and herbs, and multicultural approaches to food preservation through fermentation

and pickling; this forms the basis for an exploration into food democracy. We are looking at some of the ingredients, processes and catalysts for transformation to create an active food citizenship and shake up race, class, gender and other flavours into the mix of what has been so far in the UK, largely white, middle class and privileged cooking pots and tables.

In what way is it art, research and activism?

Cooking is the most universal form of art practice and the most accessible. It is imagination and play; it indulges all your senses, seducing you with colours, textures, scents and sounds. Recipes are references and inspiration for individual and collective creativity with food (Figure 2.4).

As research, cooking is for me a method of participatory research. It creates a space for understanding food from a different worldview, and from the everyday experiences of people. Jennifer Brady put it in her article *Cooking as inquiry: A method to stir up prevailing ways of knowing food, body and identity*: 'Cooking as inquiry recognizes bodies and food as sites of knowledge and engages researchers as researcher-participants in reflexive, collaborative study that explores the ways in which the embodied self is performed relationally through food-making (Brady, 2011).

Food and our food system is a hotbed of fermentative politics in a quest for justice. Cooking is resistance. As an individual call to action, Mark Bittman states that 'Cooking real food is the most radical thing we can do. … [I]t takes power away from the giant food corporations and restaurant chains … and puts us back in control over what goes into our bodies' (Bittman, 2014).

Why did you choose this artistic methodology to work in the field of transforming food systems?

Things needed to be done differently to engage people in a meaningful way to bring about change in a system that is so deeply embedded in society, bound up with multiple oppressive structures.

> For the master's tools will never dismantle the master's house. They may allow us to temporarily beat him at his own game, but they will never enable us to bring about genuine change. Racism and homophobia are real conditions of all our lives in this place and time. I urge each one of us here to reach down into that deep place of knowledge inside herself and touch that terror and loathing of any difference that lives here. See whose face it wears. Then the personal as the political can begin to illuminate all our choices.
>
> *(Lorde, 1984)*

What kind of transformative effect are you hoping to achieve?

Food is one of the key tools for breaking the cycle of racism, discrimination and hatred towards people. It enables one or more groups of people to come together in a collective manner and share their stories, cultures and experiences with dignity and equity, valuing diversity and difference.

Are you accessing new forms of knowledge production? How?

No. I am accessing different types of knowledge that have not been acknowledged or valued in Western society. Valuing emotions and emotional/spiritual connections as knowledge and not something to be dismissed is a reclamation of other paradigms around food, and the natural world.

How does it create new forms of public dialogue and debate?

Cooking has opened a much needed social forum for ordinary people, reclaiming the heart and soul from the cult of celebrity, and so-called experts. Marginalised, unheard voices are able to participate in the dialogues around food, bringing very different experiences and understandings, enabling disruptions and shifts in the current paradigm.

Have you already seen transformations happening in the participants and food system? How?

I have seen people change their behaviour, change their shopping and eating habits, to learn more about where their food comes from, or to grow their

FIGURE 2.4 Deirdre Woods recipe for Haitian Pikliz. *Source:* Deirdre Woods (used with permission).

own. Yet others have become more involved in food activism and other food-connected activisms such as environment and climate change, to poverty and hunger.

Do you feel you are becoming a performer and that you make resistance and transformation consumable through your work?
My authenticity, in how I connect with people, is key to my work. Performance for me engenders notions of pandering to the shallow aspects of social connection, like the cult of celebrity. I am offering a tangible experience that is easy to relate to, enables others to recognise their own power and how to use it.

What aspects of feminism are important for your work?
Cooking itself inhabits a contested gendered arena in and out of the home; however, the power relationships within the food system and around food and cooking are intersectional. As a working class, disabled, woman of colour who navigates the world through these intersections, I cannot just abstract feminism as the only lens in which to embody, analyse, or practise cooking.

Part 3: Tying a bow/braiding a plait

As our string figures come to a close, we wish to reflect on the values they contribute to thinking in urban agroecological terms. In summary, each project in our string figure includes opportunities for dialogue and some provide ways in which to view scientific knowledge as an experience rather than as a result. Therefore, the outcome is not an answer but rather the act of opening up in order to bring up inconvenient questions, new forms of reflections and research modes.

As most artistic processes, especially those in collectives, involve numerous rounds of reflection, it becomes difficult to question results because works are based on subjective and situated experiences. Each contribution examines or showcases varied models for knowledge development based on a *situated* understanding and production. For *Grey, Haide, Orrù, Ujuaje* and *Van Dyck*, agroecological discourses are put into a performative construct that draws in the audience as activated and participating agents. Limitations to this are about rendering the performance not purely as an aesthetic task but one that includes the audience in the task to bring up new insights, discourses and shifts in behaviour. All that remains afterwards is equally as critical as the process, enactment and experience itself. For *Halder* and *Solomon*, the task lies in forming a commons for collective knowledge-making. The challenge to this may be in what occurs to the group efforts once the artist steps away.

All contributors agree that there is need for new ways of enquiring to discuss difficult political and ethical matters. This opens up another dilemma: working with actors beyond urban citizens belonging to a privileged class with spare time to participate, and involving producers, farmers, planners, politicians and civil society, who need to be included in the transformation. There is a potential for agroecology to open up spaces in this process for the inclusion of the ones marginalised and overlooked; for example the women who cook every day, the passed down generational recipes and the community elders.

To enrich the agroecology discourse, other models of knowledge development are needed, giving a re-envisioned place for food and feminist practices into the discussion. This discourse must step away from representing 'food' simply as a statistic so that it can become a working material and mode for creating new attentiveness into the way we relate to food and to urban space.

Notes

1 A *string figure*, also called '*cat's cradle*', is a game that can be played by two or more people, where you use your fingers to manipulate a string around, on and in between players' hands.
2 Each participant will be indicated in italics inside parentheses inside the text, i.e., (*Haide, Orrù*).
3 *Urbaniahoeve* is a Dutch word for 'the city as our farmyard'.
4 Artistic research is relatively new to fields of urban food systems, geography, urban planning, and social science but has an accepted underpinning in Scandinavian contexts, present in academic contexts since the 2000s (Biggs and Karlsson, 2011, Hannula et al., 2005).
5 Haraway (2003) refers to human–more-than-human relationships as '*naturecultures*' in her Companion Species Manifesto, allowing us to think through these relationships from a horizontal perspective.
6 The notion of poethics (Camello Pinilla, 2016) helps describe new relational encounters: contact, listening, reflection, and being in relation to others, human and non-human, to foster a creative, embedded and collective enactment (Orrù 2017a, Camello Pinilla, 2016, p.41).
7 One reaction is to use collective financing methods, such as crowdfunding, or producing outlets, such as exhibitions or publications.
8 https://threeacresandacow.co.uk/reviews/ [Accessed 9 June 2018].
9 Films accessible at: www.communitygarden.de and www.saatgutkampagne.org [Accessed 17 May 2019].
10 Text is accompanied by a film essay (Orrù, 2017b).
11 Butoh is a Japanese dance form that trains embodied understandings between nature and space.
12 This is a spatial metaphor for a synthesis between *nature* and *culture* (Haraway 2003).
13 Other performative material properties include slime, bacteria, flower, salad and stone deriving natural worlds such as bacteria, light, tree, bee, or stone world (SU-EN et al., 2003).
14 Solomon artwork – w/the soil-building community of the food forest ecosystem, 2011–2019. *Topsoil of the Urbaniahoeve Food Forest Ecosystem in Amsterdam Noord.*
15 Solomon artwork – 2016 – *Soil Portraits #01 – 80* chromatography and topsoil production.
16 Solomon artwork – 2013 – *FREE FOOD FOR ALL*, manifesto, maps, drawings of Amsterdam borough van der Pek as urban food forest.
17 Solomon artwork – 2014 *Radical Observation.*
18 www.communitycentredknowledge.yolasite.com [Accessed 9 June 2018].
19 Until lions have historians or storytellers, tales of hunting will always glorify the hunter. West African Proverb. (http://www.afriprov.org/african-proverb-of-the-month/32-2006proverbs/224-april-2006-proverb-quntil-the-lion-has-his-or-her-own-storyteller-the-hunter-will-always-have-the-best-part-of-the-storyq-ewe-mina-benin-ghana-and-togo-.html [Accessed 9 June 2018].

20 The lecture in motion 'Reclaim Research' (https://conferences-gesticulees.be [Accessed 9. June 2018]) was developed mainly during training with the popular education group la Volte.
21 www.fieldliberation.org [Accessed 9 June 2018].
22 Free online version of Boaventura de Sousa Santos' anthology *Another Knowledge Is Possible: Beyond Northern Epistemologies* (2007) www.boaventuradesousasantos.pt [Accessed 9 June 2018].

References

Ahmed, S. (2006) *Queer Phenomenology: Orientations, Objects, Others.* Duke University Press, Durham.

Ares, P. and Risler, J. (2016) *Manual of Collective Mapping. Critical Cartographic Resources for Territorial Processes of Collaborative Creation.* [online] Issuu: https://issuu.com/iconoclasistas/docs/manual_mapping_ingles

Ayala, J. and Zaal, M. (2016) 'Poetics of Justice: Using Art as Action and Analysis in Participatory Action Research', *Networks: An Online Journal for Teacher Research*, 18(1), p.2.

Bennett, J. (2010) *Vibrant Matter: A Political Ecology of Things.* Duke University Press, Durham.

Biggs, M. and Karlsson, H. (eds) (2011) *The Routledge Companion to Research in the Arts.* Routledge, London.

Bittman, M. (2014) interview with Wilson, A.L. 'Mark Bittman Explains Why Cooking at Home Is the "Most Radical Thing" We Can Do', *Edible Manhattan* [online] [Accessed 9 June 2018].

Brady, J. (2011) 'Cooking as Inquiry: A Method to Stir Up Prevailing Ways of Knowing Food, Body, and Identity', *International Journal of Qualitative Methods*, 10(4), pp.321–334.

Camello Pinilla, S. M. (2016) *(Po)ethical Indigenous Language Practices: Redefining Revitalisation and Challenging Epistemic Colonial Violence in Colombia.* PhD Thesis, Goldsmiths, University of London.

Carolan, M. S. (2009) 'I Do Therefore There Is': Enlivening Socioenvironmental Theory', *Environmental Politics*, 18(1), pp.1–17.

de Sousa Santos, B. (ed.) (2007) *Another Knowledge is Possible: Beyond Northern Epistemologies.* Verso, London.

Diamond, E. (1996) 'Introduction', in Diamond, E. (ed.) *Performance & Cultural Politics.* Routledge, London, pp.1–12.

Dyrssen, C. (2015) 'Artistic Research - A Subject Overview 2014', in Lind, T. (ed.) *From Arts College to University: Artistic Research Yearbook 2015.* Swedish Research Council, Stockholm, pp.20–27.

Fischer-Lichte, E. (2009) 'Performative Spaces and Imagined Spaces: How Bodily Movement Sets the Imagination in Motion', in Huppauf, B. and Wulf, C. (eds) *Dynamics and Performativity of Imagination: The Image between the Visible and the Invisible.* Routledge, New York, pp.178–187.

Fox, M. (2015) 'Embodied Methodologies, Participation, and the Art of Research', *Social and Personality Psychology Compass*, 9(7), pp.321–332.

Freire, P. (1970) *Pedagogy of the Oppressed.* Continuum, New York.

Frichot, H., Preston, J., Spooner, M., Pickersgill, S., Kovar, Z., Hann, C. and Evans, M. (2012) 'An Antipodean Imaginary for Architecture + Philosophy: Ficto-Critical Approaches to Design Practice Research', *Footprint*, 1011, pp.69–96.

Fukuoka, M. (1978) *The One-Straw Revolution: An Introduction to Natural Farming*. Rodale Press, Emmaus.

Haide, E. (2003–2018) *Another World is Plantable! Parts 1–6, Documentaries Series on Community Gardens around the World*. Pullach, www.eine-andere-welt-ist-pflanzbar.de [Accessed 1 February 2019].

Haide, E. (forthcoming) *Queer Gardening*. Pullach, www.eine-andere-welt-ist-pflanzbar.de [Accessed 1 February 2019].

Halder, S. (2018) *Gemeinsam die Hände dreckig machen - Aktionsforschungen im aktivistischen Kontext urbaner Gärten und kollektiver Kartierungen*. transcript, Bielefeld.

Hannula, M., Suoranta, J. and Vadén, T. (2005) *Artistic Research – Theories, Methods, and Practices*. Academy of Fine Arts, Helsinki and University of Gothenburg, ArtMonitor.

Haraway, D. J. (1988) 'Situated Knowledges: The Science Question in Feminism and the Privilege of Partial Perspective', *Feminist Studies*, 14(3), pp.575–599.

Haraway, D. J. (2003) *The Companion Species Manifesto: Dogs, People, and Significant Otherness*. Prickly Paradigm Press, Chicago.

Haraway, D. J. (2012) *SF: Speculative Fabulation and String Figures/SF: Spekulative Fabulation und String-Figuren*. Hatje CantzVerlag, Berlin, pp.1–18.

Haraway, D. J. (2016) *Staying with the Trouble: Making Kin in the Chthulucene*. Duke University Press, Durham.

Hauptmann, D. (2006) *The Body in Architecture*. 010 Publisher, Rotterdam.

Jasanoff, S. (ed.) (2004) *States of Knowledge: The Co-production of Science and the Social Order*. Routledge, New York.

Latour, B. (1998) 'From the World of Science to the World of Research', *Science*, 280(5361), pp.208–209.

Lorde, A. (1984) *Sister Outsider: Essays and Speeches*. Crossing Press, Berkeley.

Loveless, N. S. (2012) 'Practice in the Flesh of Theory: Art, Research, and the Fine Arts PhD', *Canadian Journal of Communication*, 37(1), pp.93–108.

Marchand, T. H. (eds) (2011) *Making Knowledge: Explorations of the Indissoluble Relation between Mind, Body and Environment*. John Wiley & Sons, Hoboken.

Mortimer-Sandilands, C. and Erickson, B. (eds) (2010) *Queer Ecologies: Sex, Nature, Politics, Desire*. Indiana University Press, Bloomington.

Orrù, A. M. (2017a) *Wild Poethics – Exploring Relational and Embodied Practices in Urban-Making*. PhD Thesis, Chalmers University of Technology.

Orrù, A. M. (2017b) Organoleptic Interfaces [essay film], Sweden, https://www.youtube.com/watch?v=8CIG6ElaCG4 [Accessed 1 July 2018].

Puig de la Bellacasa, M. (2015) 'Making Time for Soil: Technoscientific Futurity and the Pace of Care', *Social Studies of Science*, 45(5), pp.691–716.

Solomon, D. (2012) 'Urbaniahoeve: Expanded Urban Agriculture,' in Viljoen, A. and Wiskerke, J. S. C. (eds) *Sustainable Food Planning: Evolving Theory and Practice*. Wageningen, NL: Wageningen Academic Publishers, pp.365–382.

Solomon, D. and Nevejan, C. I. M. (2019) 'Soil in the City: The Socio-Environmental Substrate', in Toland, A., Stratton Noller, J. and Wessolek, G. (eds) *Field to Palette: Dialogues on Soil and Art in the Anthropocene*. Taylor & Francis Group, CRC Press, Boca Raton, pp.605–624.

Solomon, D. and van den Berg, M. (2014) 'Urbaniahoeve: Event-Based Practice and Urban Agriculture', in Viljoen, A. and Bohn, K. (eds) *Second Nature Urban Agriculture: Designing Productive Cities*. Routledge, New York, pp.200–205.

Spaid, S. (2017) *Ecovention Europe: Art to Transform Ecologies, 1957–2017*. Museum De Domijnen – Hedendaagse Kunst, Sittard.

SU-EN, Sandberg, M. and Kennedy, G. (2003) *Butoh - Body and the World*. Van Rye Publishing, Ann Arbor.

Toland, A., Stratton Noller, J. and Wessolek, G. (eds) (2018) *Field to Palette: Dialogues on Soil and Art in the Anthropocene*. Taylor & Francis Group, CRC Press, Boca Raton.

Tsing, A. L. (2017) *The Mushroom at the End of the World: On the Possibility of Life in Capitalist Ruins*. Princeton University Press, Princeton.

3

COMMONS AND COMMONING FOR A JUST AGROECOLOGICAL TRANSITION

The importance of de-colonising and de-commodifying our food systems

Tomaso Ferrando, Priscilla Claeys, Dagmar Diesner, Jose Luis Vivero-Pol and Deirdre Woods

> "Àjó je dun lò mu işe şe şe"[1]
>
> Yoruba proverb

Section I – Exploring food systems through post-commodity and post-colonial lenses

The history of the global food system is characterised by the transformation of the tangible and intangible commons (i.e., land, water, knowledge, seeds, work) into individual proprietary entitlements (mostly given to male owners) within political and power structures that benefitted settlers and wealthy landowners at the expense of indigenous peoples and other rural communities (Hamilton and Bankes, 2010; Federici, 2019). More specifically, when it comes to the construction of the European food systems, including both continental Europe and the UK, four major waves of enclosure and commodification of commons can be identified:

- the Early Medieval enclosures of land and femininity (Federici, 2004);
- the colonial period that added millions of hectares of land in America, Africa, and Asia to the land used to feed Europeans (Friedmann and McMichael, 1989);
- the early days of modern capitalism and industrialisation, with the alienation of labour and the dismissal of the care economy undertaken by women (from that very moment considered as non-productive and a private household issue) (Moore, 2015; Federici, 2019);
- the neoliberal phase of financial capitalism (in the last quarter of the 20th century), where traditional factors of production (work, land, natural

resources, knowledge) were increasingly de-localised, monopolised and downplayed in relation to the hegemonic factor: the fictitious capital that is exclusively based on trust and convincing narratives by economic actors (Ferrando, 2018; Schiller, 2019).

For centuries, European cities have been fed with the products of plantations and slavery, beneficiaries of long-distance trading and often in direct competition with cities located a few miles away from the origin of the food (Haraway, 2015). The construction of contemporary capitalism was accompanied by the normalisation of the idea that food is nothing but a commodity (Vivero-Pol, 2019).[2] Through the lenses of World System Theory (Wallerstein, 1974; Arrighi, 1994), the establishment of the European food system is intrinsically connected with the dismantlement of collective forms of proprietary regimes, both in the Global North and the Global South, the erasure of collective governing mechanisms and commons infrastructures, and the subordination of soil, territory, and lives to the needs of European citizens and productivism (Sassen, 2014; Mattei and Nader, 2008; Linebaugh, 2008). Non-capitalist economies and non-individualist forms of living were neglected or directly attacked as backward, inefficient, or pre-modern; therefore, the primary goal of those actions was to replace non-market-based economies with market-driven ones.

The present of the European urban food system is thus rooted in the history of the colonial, the masculine, and the commodified, intertwined with the transformation of food, gender and nature into cheap commodities (Patel and Moore, 2017). Contemporary agribusiness began forging its political and economic relationships during the colonial period. The East India Company was filling the markets of London with its products and the wallets of its investors, while starving its workers and the rest of the people in the subcontinent (Chakraborty, 2014). Land and social relationships in the Global South were subordinated to the colonisers' vision of the patriarchal society, disempowering women in the South by enclosing the commons on which they depended (Federici, 2004), whereas the products of the colonised land were used to supply markets in the Global North, where women were increasingly relegated to the kitchen without any social or financial recognition of their roles.

The objective of this chapter is to contribute to the development of a multi-pronged approach to food systems' transformation in Europe, one that resists the capitalist model (De Angelis, 2017), recognises the colonial, patriarchal, and capitalist roots of the local and international food chains that feed Europe, and promotes alternative forms of getting together, sharing, and co-producing with nature. Indeed, a redefinition of the contemporary European city must go hand in hand with the de-commodification and de-colonisation of its past and present. In our view, a historical, political, and ecological understanding of European cities as the beneficiaries of uneven development provides epistemological and methodological tools to bridge the gap between cities and the countryside. This understanding supports agroecological urbanism as a technical and political reaction to the food-disabling urban landscapes that have separated urban dwellers from the food they eat, from the nature that they exploit, and from the living condition of the people that make food possible everywhere in the world (Deh-Tor, 2017). Indeed, it contributes to exploring a new paradigm for urbanisation, adding to the principles of solidarity, mutual learning, inter-species exchanges, environmental stewardship, food sovereignty, and people's resourcefulness (Deh-Tor, 2017). In addition, given the central role that food plays in the provision of care, it puts at the center the perspective of feminist theorists and activists who have long challenged the link between masculine enclosures of women-tapped commons and the exploitation of reproductive labour as founding pillars of the capitalist development in the past centuries (Federici, 2019; Patel and Moore, 2017).

We discuss three case studies that represent different 'loci of contestation' of the absolute commodification of food. Taken together, these cases show how the political, imaginative, and organisational power of commons and commoning can bridge the urban-rural divide, and contribute to the convergence of various movements, including agroecological urbanism (Tornaghi, 2017) and food sovereignty (Rivera-Ferre et al., 2014). These examples also point to the importance of adopting a political vision of commons and 'commoning' as intersectional antidotes to co-optation in the food movement. Through a combination of theory and practice, history and imagination, empowerment and de-commodification, this chapter brings to the forefront those dimensions of food that cannot be monetised and defined in mere monetary terms. We call for policies grounded in valuations of food that do not follow mainstream scholarly economics[3] but listen to and learn from non-Western non-economic epistemologies and promote non-heteropatriarchal visions of the food systems. In the context of the edited volume, we also believe that the paradigms of commons, commoning (Dardot and Laval, 2015; Ferrando and Vivero-Pol, 2017), and commons-based food systems may support agroecological urbanism in defying the set of capitalist social relationships (including the disempowerment of food-producing women by curtailing their control over food-producing commons and their seclusion within the household walls) that divide local communities (Dalla Costa, 2007).

In the next section, we introduce the main ideas that underlie our collective reflection, i.e., commons and commons-based food systems, commoning, and food as commons. Section III then takes us to Geneva, where peasant organisations mobilised international human rights law to obtain the recognition of their collective rights to and relationship with nature, in the framework of food sovereignty. Section IV moves the setting to London, where organized urban dwellers are fighting for a food system that is bottom-up, inclusive, anti-colonial, anti-patriarchal, and constructed around food sovereignty and commoning. Section V takes us to the Italian city of Bologna, where individuals and associations are deploying the paradigm and practices of the food system as a commons to resist an economic and social framework based on expulsion, subordination, and domination (Sassen, 2014).

The cases presented here have been documented from within by those of us who, as scholar-activists, share a commitment to supporting various social movement struggles that seek to advance and protect commons and commons-based food systems, commoning, and food as commons. These cases offer a unique insight into the dynamics, tensions, and aspirations of different groups that, vocally or quietly (Visser et al., 2015), are experimenting with the ideas of food and the food systems as commons in various locales.

Section II – Commons and commons-based food systems, commoning, and food as commons

Being a de-colonial and feminist group of authors, we cannot but recognise the plurality of meanings and institutional arrangements the commons have in contexts characterised by different epistemologies (de Sousa Santos, 2018; Kothari et al., 2019) and praxis (Mattei, 2013). We are also aware that the notions of 'commons' and 'commoning' may not resonate with all people, communities, and individuals, and that the idea of the common good was already appropriated in the past in order to justify Western moral superiority and legitimise the occupation of land. In addition, we believe there is much to learn from the ways in which commons and 'commoning' are lived, experienced, and reproduced on a daily basis around the world, with important differences between urban and rural settings (not only in the Global South and not only in rural areas).[4]

There is not one 'correct' interpretation of the commons, or one way of translating the idea into practice. Rather, authors and disciplines have been elaborating very different conceptions, spanning from pure economic considerations of common-pooled resources (rival but difficult to exclude) to the understanding of the commons as the catalyst of anti-capitalist vindications. Commons have been described based on the inner nature of the good, the proprietary regime, or the governing mechanisms.

Overall, we agree that the commons are not defined by the ontological properties intrinsic to the goods, but rather by collective decisions (*ergo* phenomenological) that are context-specific and highly conditioned by the contextual and material co-construction of nature and society as an ecological *unum*. Moreover, we all share a vision of the commons that is not static and definite, but that considers the commons as constantly dialoguing with 'commoning', and therefore being always redefined by collective action. It is 'commoning' together that confers to a material and non-material resource its commons consideration (Dardot and Laval, 2015).

We use the term commons-based food systems to designate the natural resources and practices that underlie and enable the collective and democratic management of the material and immaterial components that are essential to the establishment of fair, sustainable, resilient, and self-governed food systems (Pettenati et al., 2019; Maughan and Ferrando, 2019). Contrary to the industrial and capitalist food system in which resource accumulation and exploitation for profit are the norm, a commons-based food system revolves around collective governance, rational utilisation of natural resources (considering the livelihood of future generations), and a fair distribution of revenues and food products (Vivero-Pol et al., 2019).

Those among us who promote the idea that food could/should be governed as a commons (as a complementarity to 'commoning') see food as a life enabler and a cultural cornerstone, a resource with multiple meanings and different valuations for societies and individuals (Wall, 2014; Szymanski, 2016). From this point of view, food shapes morals and norms, triggers enjoyment and social life, substantiates art and culture (gastronomy), affects traditions and identity, relates to animal ethics and determines, and is shaped by power and control. These multiple and relevant meanings cannot be reduced to the one of tradable good, and the value of food cannot be fully expressed by its price in the market. As a result, food cannot be reduced to a commodity[5] (Vivero-Pol, 2017).

The examples discussed in the next three sections express our different visions of commons, commoning, and commons-based or commons-generating food systems but are kept together by our shared interest in learning from and supporting the struggles and practices associated with these notions and with decommodified, anti-colonial, and anti-patriarchal visions of the food systems. We start with the experience of the transnational agrarian movement La Via Campesina (LVC), which struggled for, participated in the drafting of, and achieved the adoption of a United Nations Declaration on the Rights of Peasants (hereafter UNDROP)[6] and other people working in rural areas by the UN General Assembly in December 2018. The adoption of UNDROP marks the culmination of a 17-year process aimed at obtaining international recognition of peasants' collective human rights to land, seeds, and the means of food production. It articulates a set of anti-capitalist and anti-imperial values, assumptions, and objectives that are radically opposed to that of enclosure, commodification, and unsustainable exploitation and control of nature.

There is no doubt that international public law is embedded in colonial roots and central to the legitimation of violence through the recognition of boundaries, frontiers, and the absolute sovereignty of states over people (in particular indigenous people) and the planet (Anghie and Chimni, 2003). Yet the case of UNDROP is interesting precisely because peasant movements and their allies identified international human rights law as a space of power and political contestation. Advancing the paradigm of collective human rights, associated with that of the commons, these movements used the UNDROP process to open up dialogue and much needed societal debate on the peasantry, but also on the future of food and humanity.

Section III – Commons in UNDROP[7]

Approximately 2.5 billion men and women from indigenous, peasant and other rural communities worldwide are estimated to depend on lands managed through customary, community-based tenure systems. These lands would account for over a quarter of the world's land surface, intersect about 40% of all terrestrial protected areas and ecologically intact landscapes (Garnett et al., 2018), and produce more than half of the food consumed every day in the world (Kay, 2016). However, these communities have formally recognised land rights over only one-fifth of these territories (RRI, 2015). Lack of recognition of their customary rights and persisting marginalisation, coupled with biased approaches towards collective forms of land ownership and use, and the gradual erosion of their traditional institutions, in many cases entails a great vulnerability of commons to appropriation by the state and private actors (Thornberry and Viljoen, 2009).

Over the past decades, indigenous peoples, peasants and other people working in rural areas organised at the transnational level, have brought their struggles to retain control over their lives, livelihoods, lands, and territories to the United Nations, framing their claims within a human rights discourse (Errico, 2017; Claeys, 2015). Their claims have triggered the drafting of new international human rights instruments stemming from discrimination-related considerations. The UN Declaration on the Rights of Indigenous Peoples was adopted in 2007 by the UN General Assembly, and the UN draft Declaration on the Rights of Peasants and other people working in rural areas was adopted in September 2018 by the UN Human Rights Council (HRC), and three months later by the UN General Assembly. At the core of these instruments are provisions that recognise and seek to protect, to varying degrees, their collective rights to lands and natural resources, including commons that are a source of food.

The process of elaboration of UNDROP was initiated in 2012 under the leadership of Bolivia, at the request of the transnational agrarian movement La Via Campesina. The Declaration, the initial draft of which was elaborated by LVC in the late 1990s, constitutes a direct reaction against the enclosure of rural commons, what Borras and Franco (2012) have called a peoples' counter-enclosure. Seizing the political opportunity created by the global food crisis of 2007–2008, which highlighted new waves of land and green grabbing and the challenges facing small-scale producers in their access to land, LVC succeeded in putting its demand for new peasants' rights on the agenda of the Human Rights Council.

At the first session of the Open-Ended Intergovernmental Working Group established in 2012, states and observers (including LVC activists, allies, and technical experts) used the draft elaborated by LVC as a basis for discussion (UN, 2012). This draft adopted a comprehensive approach to human rights that goes beyond the individual entitlement and a holistic understanding of the interaction between food production, transformation, and consumption. It included the main demands of peasants and other groups of food producers, namely the right to land, the right to seeds, the right to biodiversity, and the right to food sovereignty. It placed these new rights (not yet formally recognised as such in the international system of human rights) within the broader framework of already recognised civil, political, economic, social, and cultural rights, with a special emphasis on the right to organise, access to justice, and the criminalisation of human rights defenders.

This initial draft contained no explicit reference to the commons, but it insisted on the collective dimensions of the right to land and natural resources in the broader context of a food and economic system where human beings and nature are intrinsically interconnected. Article 4, on the right to land, listed unused land, forests, fishing grounds, and territories as important areas where peasants rear livestock, hunt, gather, fish, or toil. Article 4 emphasised peasants' rights to manage, conserve, benefit, toil, and produce. It further included the right to benefit from land reform, with explicit references to the prohibition of land evictions and the need for land regulation measures such as land ceilings and the prohibition of *latifundia* (large-scale exploitations), to address widespread issues of landlessness and displacements. As such, peasant activists demanded that their individual and collective rights to land and their special and direct relationship to land and nature (Article 1.1 on the definition of peasants) be recognised, respected, and protected from private and state interference, while calling on the state to regulate land markets and redistribute land.

Between 2012 and 2018, five sessions of negotiations took place at the HRC under the chairmanship of Bolivia, leading to successive drafts of the Declaration. In the final draft, which was adopted in September 2018, Article 17 of the Declaration contains the main provision concerning land and natural resources, and recognises an individual and collective human right to land; it also contains an explicit reference to the 'commons' (UN, 2018). Specifically, the article provides for the legal recognition of existing customary land tenure rights and establishes an obligation for states to "recognize and protect the natural commons with their related systems of collective use and management" (Article 17.3).

The text used in this article builds on the language used in paragraph 8.3 of the Voluntary Guidelines on the Responsible Governance of Tenure of Land, Fisheries and Forests in the Context of National Food Security (VGGT) unanimously adopted by the Committee on World Food Security (CFS) in May 2012 (CFS, 2012). Paragraph 8.2 of the VGGT states that "the legitimate tenure rights of individuals and communities, including where applicable those with customary tenure systems, should be recognized, respected and protected". Paragraph 8.3 of the VGGT notes the existence of "publicly-owned land, fisheries and forests that are collectively used and managed (in some national contexts referred to as commons)," and calls on states to "recognize and protect such publicly-owned land, fisheries and forests and their related systems of collective use and management, including in processes of allocation by the State".

The inclusion of specific provisions calling for the recognition and protection of the commons in these two international instruments, within the broader context of the emergence of an individual and collective right to land, could significantly contribute to the transition to ecological and integrated food systems. This would be based on the access to and reproduction of the commons (i.e., coastal fisheries, hunting grounds, forest foods) rather than their exploitation and subordination to the needs of consumption. Indeed, rural households all over the world depend, to a significant extent, on the commons and are particularly threatened by their enclosure. In the future, the Declaration could directly support the protection of the commons because it contains provisions concerning the overall 'enabling' environment concerning commons, touching on key aspects such as participation in policy-making and trade-related issues, among many others. However, embedding the recognition and protection of the commons within the human rights framework raises interesting and complex questions when it comes to the interactions between commons, state, and market.

UNDROP is grounded on the 'respect, protect, and fulfil' framework typical of human rights instruments, which points to specific sets of actions on the part of the state. All its provisions are therefore framed around this scheme, according to which the state shall:

(1) refrain from interfering or curtailing the enjoyment of the rights concerned;
(2) protect individuals and groups against abuses by third parties, including business enterprises;
(3) take positive action to ensure the enjoyment of these rights by facilitating or providing the conditions necessary for this (see, for example, UN, 2009).

More specifically, states are expected to recognise and protect the customary tenure systems of peasant communities, including commons, and support their long-term viability. They should play a redistributive role if there is unequal access to land, natural resources, and means of production, paving the way for the potential establishment of new commons within the context of redistributive agrarian reforms. They should also devolve authority and power to local communities, recognising these communities as co-managers of natural resources and their socio-ecological setting, while ensuring that commons are self-governed in an inclusive, accountable, and sustainable manner. Similarly, the VGGT calls on states to secure and implement legitimate tenure rights (including to commons), including by devolving authority and responsibility to govern natural resources to the local level.

While certain approaches to the commons see commons as key innovative and transformative tools that would help food systems move beyond the state and the market (Bollier and Helfrich, 2015), UNDROP forces researchers and activists to consider the possible tensions and synergies that could be found between the human rights approach that relies on the actions of the public authorities (including city, counties, regions, etc.) and the commons approach, where self-regulated actions of the collectivities are often emerging outside of the state-market dualism. In addition, while the recognition of collective human rights that is at the heart of UNDROP is a clear attempt at de-colonising the international human rights system (Claeys, 2019), the implementation of the Declaration will raise complex questions relating to the fulfilment of individual rights within communal rights systems, and the settlement of conflicts between various users of the land at the local level.

When transposed from the international level to the localism of European cities, commons and commoning need activists, politicians, and academics to think historically and consider the present, both through the past and the future. The case of the UK presented in the next section is therefore an attempt to look at the way in which the food sovereignty movement could be enriched and strengthened by the adoption of an intersectional and historically strong engagement with the colonial ties of the current food system and its repercussions in terms of social and environmental injustices.

Section IV – The missing post-colonial approach in the UK Food Sovereignty Movement

We believe that a critical and politically oriented exploration of the link between food and the commons triggers fundamental questions relating to colonialism, post-colonialism, commodification, and social justice. This is nowhere more evident than in the UK foodscape, a space rooted in colonial ties and reproductive of the colonial legacy. If we want to think through the lenses of global enclosures, expulsions, and appropriations, we also need to think of the mounting obesity pandemic, the rising figures of food insecure households, and the lack of power by citizens to govern their own food systems through the lenses of history and intersectionality. Therefore, it is important to ask whether the urban citizens and predominantly farmer-led UK Food Sovereignty Movement that was created in 2012, whose aim is to create a fairer and more sustainable food system in the UK (Shawki, 2015), plans to do so with or without incorporating feminist, post-colonial, and post-commodity approaches. The discussion that follows, based on the personal experience of one of the authors as a community food practitioner and activist-researcher, looks at the UK Food Sovereignty Movement as an invitation to imagine and implement practices that are fairer not only to ecology and eaters but also to marginalised communities (i.e., diaspora) and groups (i.e., women farmers).

The UK Food Sovereignty Movement was born in 2012 from a gathering of food producers, academics, and NGOs in London. Adopting six principles of the Nyéléni Declaration, a grassroots union of farmers, growers, and food workers, the Land Workers Alliance, was also formed. Since then, the UK Food Sovereignty Movement has been pushing for a radical change in the narrative and positionality of the UK food movement. With an accent on deepening diversity of policy choices, governing mechanisms, and management practices, the UK food space is shifting from being a mere receiver of top-down policies to becoming a vocal actor in crafting preferred policies. Throughout the years, the movement has aimed to improve the livelihoods of its members and create a better food system for everyone by building networks and solidarity, training, campaigning, and lobbying. At a second gathering in 2015, three key strands and working groups emerged: to develop an integrated people-led food policy for UK, to explore issues around land, and to deepen the diversity of the movement.

The great success of the movement has so far been its role as a policy disruptor. Landworkers Alliance research, lobbying, and position papers on agricultural and food-related policies have broadened the public debate in the UK,[8] as has the mobilisation of various food movement actors to collaborate and deliver an integrated national food and agricultural policy proposal 'A People's Food Policy' (Butterly and Fitzpatrick, 2017). This has contributed to an increased cohesion among different movements and the construction of a coherent background and way forward. Likewise, the 'Land for What' conference (November 2016) resulted in the emerging Land Justice Network,[9] consisting of academics, housing, food, and land reform activists; this network is re-setting the terms of debate on land proprietary regimes, land uses, and collective decisions on land governance.

Elsewhere, the development of the food sovereignty movement in Latin America and Sub-Saharan Africa has been entwined with anti-colonial struggles. On the contrary, there is a sense that the acknowledgement of Britain as the 'mother country' of the British Commonwealth, the seat of the first truly global empire, the birthplace of the industrial revolution, and the epicentre of global finance (Akala, 2018: 2), is severely lacking in the discourses and interactions that take place in the UK. The work of food regime theorists comes in support here (McMichael, 2005; Bernstein, 2016). The commodification of land, food, and labour; the displacement of peoples; and the relegation of women to the kitchen are inextricably linked to the construction of the first global food regime and, therefore, to the current setting. Moreover, these processes are not a mere memory of a remote past. Most of the land appropriated during colonial times was never redistributed and is still connected with the global food system. In addition, evictions and expulsions keep happening; the full emancipation of women is still far from materialising; and the appropriation of water, seeds, culture, and knowledge is still a reality around the world. Yet, those practices do not form the core claims of the UK Food Sovereignty Movement, mostly because they happen somewhere far from the UK, in distant places and those ill-management practices usually happen to "the Others" (Sibley, 1995; Said, 1978; Fanon, 1963). Moreover, UK research on food-related issues is still linked to European imperialism and colonialism with only a few examples of decolonial approaches to food justice (Bradley and Herrera, 2016) and colour-focused reflexivity "in the practice of making power visible at all levels" (Batliwala, 2010: 18). Other ways of knowing and understanding food and food systems have not found enabling spaces in the Westernised epistemology of food narratives, mirroring the marginalisation found in other areas of knowledge (see de Sousa Santos, 2014).

For the de-commodification of food and the construction of a socially and environmentally just food system, the intersections of social power and oppression tied to colonialism and post-colonial relationships must necessarily be a central theme in the food transformation discourse and practice not only in both the Global North and the Global South but also across the Global South and the Global North. These framings can enable space for a more realistic politics "to make new ideas, concepts and associations" (Light et al., 2009: 41) with food considered as a commons as one element of this. There is an urgent need to put considerations of Britain's colonial past and present ramifications at the centre of the national and global food politics discourse, collectively formulating an intersectional and historically rooted vision of the future of food. No agroecological city should be oblivious of its past and the contemporary reproduction of its legacy.

Some seeds have already been sown, but they must be nurtured. In 2016, a diverse delegation from the UK participated in the 2nd European Food Sovereignty Forum[10] (one of the authors was part of that delegation). A paper on "*Decolonisation and Food Sovereignty in Europe: Thoughts from the edges*" (Mama and Anderson, 2016) was presented and raised questions about colonialism, post-colonial inequalities that are still operational, and the narrative and practice of food sovereignty in the Global North. As part of the post-forum reflections, critical race theory, food justice, and the commons were considered as three pivotal elements to advance the de-colonisation of food systems. A more fundamental question also emerged, namely '*How is the food sovereignty movement in the Global North, particularly in Europe and the UK, reinforcing the very structural oppressions that it claims to challenge with food justice and democracy*?'.

A first consideration to this question, disturbing and ironic, is that the movement is not driven by, or deeply inclusive of, those from diaspora communities whose heritage is from those colonised/de-colonised countries where food sovereignty emerged. The UK Food Sovereignty Movement is mostly Global Northern-focused and farmer-led, with less urban participation and weak considerations of Global Southern diaspora communities. What then are the barriers, perceived or power-related, that Western food sovereignty advocates and the diaspora communities hold? Does the movement perhaps mirror Western societal divisions and oppressions?

In that sense, the daily experience within the wider UK food movement suggests that it is one of "race and class-based disparities" (Cohen et al., 2012: 81) and white privilege (Slocum, 2007). It is also gendered with the over-representation of white male leadership (Moyles, 2015; NBFJA, 2017) often reinforcing existing inequalities (Bradley and Herrera, 2016). Feminist and queer spaces within the UK Food Sovereignty Movement are challenging this predominance of white male European voices within the food system, although the marginalisation of diaspora voices from colonised/de-colonised countries is still rather obvious. Are black and ethnic minority groups so traumatised by the violence of enslavement, indentureship, and generational impoverishment that there is no desire to engage or participate in these alternative food movements (Harper, 2016)? Is the disconnection and translocation from native land and the cultural aspects of food production, transformation, and consumption preventing BAME groups participating in re-drafting food politics? Or are these spaces seen as unsafe, unwelcoming, and inherently racist?

The Global North Food Sovereignty Movement leans strongly towards 'the local' and often excludes 'the Other'. Perhaps unintentionally, the extreme food localism erases immigrants' and diaspora's contributions to food systems (Counihan, 2016). Yet, how can we think of potatoes, beans, tomatoes, etc., without considering the violent transfer of those foods and other colonial crops (and the water, soil, labour, and knowledge embodied in it) from Africa, Latin America, and Asia? Sweat, blood, and land from all over the world contributed to the diversity of the local food systems that the food sovereignty movement is trying to strengthen and scale up. The coloniality of power, knowledge, and of 'being' needs to be explored to truly understand the continued impact of historical violence (Mignolo, 2011). Place-based food systems need to recognise the geographical complexity of the local, and shed light on the historical and present intersectionality of the food system; it also needs to work for the inclusivity of the different cultures and food narratives coexisting in the same places. How can the food sovereignty movement promote the reclamation of indigenous and cultural knowledge in the UK (a multi-verse of people from different countries and cultures) with a monocultural discourse dominated by Anglo-Saxon affluent white males?

We believe that the combination of food sovereignty, agroecology, and commoning is a tool for an 'anticolonial unsettling the ongoing coloniality of power' (Wynter, 2003).[11] Decolonialism is the right to self-determination and the basis of a bottom-up, empowered, and emancipatory engagement with the food system. Agroecology is more than sustainable agriculture (practices and techniques); it is, first and foremost, a decolonial practice. Commons-based food systems and agroecological practices will reclaim and reassert the social, cultural, and spiritual aspects of food, either at the local level or by different food cultures that currently coexist at local levels. In that sense, valuing food as a commons and commoning as the practice of realising equitable and ecological food systems highlights the relevant bonds that food triggers in family and at the community level. This can be found in some diaspora communities.

Decolonial practices and narratives seek to shift power and the distribution of resources, not just as redistribution or reparations but by shifting the centre of those relationships: the imperial city (be it London, Paris, or Brussels) has been the centre of the food system for centuries. By drawing on non-Eurocentric worldviews, decolonial and unsettling practices can help develop new/old ways of food exchange, be it with nature or humans, that are based on reciprocity, mutual legitimisation, and the essentialness of food for human survival and the cultural foundations of food systems. Commons are regarded not as fixed but autopoietic relationships between nature and humans themselves, with its resources to be used by humans. People of different cultures, ethnicities, genders, and status share a common trait: all need to eat every day and all value food for dimensions other than its price. 'Humanness' as connection, practice, and way of being can be epitomised by the common need to eat food. So, being human means having a need to eat.

Therefore, this paradigmatic disruption of the commodified must go beyond the agroecological narratives and practices that challenge the neoliberal notions of private ownership, financialisation, capitalism, and market fundamentalism. A decolonial and commons-based attack against the legacies of imperialism and colonialism includes, first, the de-commodification of labour where black and brown bodies continue to be exploited as cheap labour; second, the de-commodification of means of production of food, such as seeds, land, or water (Maughan and Ferrando, 2019); and third, the de-commodification of food itself, underplaying the monetised tradable dimensions of food (that exists in any case) and re-positioning other non-monetised dimensions that are equally important for humans, such as the cultural dimensions, the right-based approaches to food, or its essentialness for our bodies (Vivero-Pol, 2019).

Together with the construction of agroecological, democratic, and regenerative food systems, we believe that the task of the Food Sovereignty Movement in the UK is to dismantle the narrative scaffoldings and structures of race, class, sex, and other oppressions that divide us. These prevent the collective agency from shifting towards a commons-based food system in the UK, where everybody has access to sufficient and adequate food to have a dignified and meaningful life. Valuing food as a commons, beyond the collective rights it may carry (a legal notion already quite disruptive), provides the underlying philosophical foundation for the system of human-nature relationships to counter the current food system paradigm based on commoditised food at the service of consumption and the neo-colonial structures of the metropole and the periphery (i.e., cash crops from the Global South, globalised open markets working only in one direction, free flow of money but not people, and land grabbing schemes in the South).

In exploring the potential of commons-based food systems as researchers, practitioners, and activists in the UK food movement, we have come across fundamental questions relating to colonialism, the construction of the contemporary urban food movements, and social justice. In particular, we felt the urgent need to redefine considerations of Britain's colonial past and present ramifications into the food politics discourse that is rapidly unfolding in the contemporary UK. This includes when (and how) it is built from the bottom-up. The way in which history and intersectionality can be integrated into the food (sovereignty) movement is not simple and is posing multiple questions that can only be answered by means of experimenting and experiencing.

All over the world, including in the Global North, there is a rise in the intersectional attempts to challenge multiple injustices that characterise the history of capitalism and the contemporary relationships between people, food, land, and socio-environmental injustices. The people behind these experiences may not use the vocabulary of the commons or think of what they are doing as 'commoning'. However, this is not important: what matters is their ability to collectively imagine and dynamically perform a food system that rejects the idea of people and nature as commodities, that is solidly rooted in the continuous and regenerative interaction between people and planet, and whose value is not only that of the exchange but is also fully aware of the multiple forms of historical and contemporary domination of people and the planet that characterise the dominant food system.

Moving on to the the third example, the last reflection of this chapter is dedicated to a new form of resistance devised by urban clients and rural food producers at the bottom of the food chain in response to the ongoing efforts by state and capital to enclose their lives through liberalised trade policies and the privatisation of food-producing resources. Drawing on the experience of Campi Aperti in Italy, we argue that the collaborative effort of rural and peri-urban producers and urban dwellers for the preservation and expansion of spaces of autonomy is crucial to the thriving of commons-based food systems involving the production, distribution, and sharing of food outside of the paradigm of the commodity. The case presented here offers insights into how a community of small farmers and critical citizens-buyers subvert some key notions of capitalist food markets, such as the sharp distinction between producers and consumers, the price-setting mechanism, the regulatory role of state agencies, and the reality of corporate governance among others.

Section V – Campi Aperti and Genuino Clandestino: de-commodifying food

The Association Campi Aperti was formed in Bologna (Emilia-Romagna region, Italy) shortly after the anti-globalisation movement gathered in Genoa in July 2001. At present, it has over 130 full-time food producers. Campi Aperti approached the struggle as a strategic operation in order to set in motion the necessary process for transformation from a subordinated and unequal position under the state and capital, with the idea of creating an ethical space in which to reproduce their livelihood as small farmers, together with new relationships to urban clients and agroecological relationships to nature. The paradigm of food sovereignty was perceived by the founders of Campi Aperti as a framework with transformative political substance, which offered sufficient latitude to be adapted to the local context in which they were operating. The founder of Campi Aperti said: "The purpose of the self-management of market exchange, food production is geared towards a market offering varied agricultural products mirroring the great local biodiversity".[12] Along the same lines, one producer that participated in the scheme promoted by Campi Aperti described the diversity of the self-management market in the following terms: "40% vegetables/fruits; 20 % cheese, yoghurt; 20% wine, beer, olive oil; 10% bread, cakes, flour; 10% herbal medicine" (Interview 2018).

One significant aspect for many producers and farmers is to earn a decent income from the sale of their products. Unlike the industrial food economy that perceives farmers, peasants, and producers as individual competitors with diverging interests, and is therefore able to dump the food prices at the farm gate to reach profit margins at the retail spot, Campi Aperti has lifted food out of the competitive markets by placing it in the solidarity economy and strengthening the social ties across all the phases of the food system. In its manifesto, it states the central value of their solidarity economy, namely to break the competitive cycle and replace it with cooperation, solidarity, equality, and sustainability (Campi Aperti, 2014). For Campi Aperti: "The solidarity economy is preferred to a market economy because it allows to establish forms of practical solidarity between consumers and producers, united by the pursuit of common objectives, such as health, the environment, and the dignity of work".[12]

The first steps away from the conflictual relationships imposed by a competitive and unequal market are the liberation of their produce from the yoke of the wholesale market distribution streams, and the emancipation of the farmers from the threat of the global-local supply chain and the constant exposure to the unstable price volatility. Instead, together with 'co-producers' (i.e., urban citizens who are critical buyers, eaters, and active participants into the system),[13] the farmers enter into a conversation about production and distribution that co-constructs the price of food. This involvement of the co-producers in the price-setting enables them to understand the ecological and labour conditions behind what they consume and what type of cultivation inputs it takes to feed them. The experience with Campi Aperti shows that the benchmark for this price-setting model guarantees a decent living for farmers and an acceptable price for consumers (because they know what they are paying for). In this sense, Campi Aperti realised that a different understanding of food and food-producing intricacies required a different type of market, one that acts as a catalyst for de-commodifying food through the forms of practical solidarity that are pursued, both through recursive practices and being institutionalised by the governance system of the association. By collectivising the decisions on price-setting, self-regulated mechanisms of labelling and production, they introduced 'commoning' practices that are useful for the community in three dimensions: provisioning of healthy and affordable food, fostering social life, and enabling peer governance (Bollier and Helfrich, 2019).

The second key element of the collaborative, equal, and sustainable food system that challenges the fetishism of the food commodity and gives visibility to all the actors of the food system is the participatory-guarantee system (PGS), a self-certified system of organic products jointly managed by the co-producers and producers. This self-managed certification system has a dual function. On the one hand, self-certifying provides an opportunity to transition towards organic food systems outside of the rigid regulatory framework of the state's certification system, the strict nature of which is such that small and medium size producers cannot afford it. However, the fact that it operates outside of the official framework means that the PGS is not a legal system that binds producers through public sanctions and deterrents, but a self-governance tool enacted by producers and co-producers. On the other hand, it acts as an internal regulator (self-regulating mechanism) for implementing agro-ecological methods and practices. It is an opportunity for food producers to certify their products without seeking state-issued or private labelling. The product's legitimacy does not derive from an impersonal state bureaucrat or a large-scale retailer, rather from the recognition and appreciation by 'co-producers' who are buying the PGS products. On the other hand, it acts as an internal regulator (self-regulating mechanism) for possible misbehaviour. This certification system has also proven to be a regulator for correcting misconducts in the long run. In this case, the producer is held accountable by taking responsibility for resolving the problems raised by Campi Aperti. If the producer refuses to deal with it, the producer has to leave. This type of self-regulation is based on a clear set of principles that form internal boundaries for well-functioning commons (Ostrom, 1990). The PGS is the instigator for three main constituents of Campi Aperti's struggle for food sovereignty: direct sale or *vendita diretta*, peasant agriculture or *agricultura contadina*, and empowerment.

The PGS is a bottom-up and collective response to the up-hill struggle with local authorities and existing pro-large business legislation that is faced by many small-scale producers of transformed foods (for example, wine makers, bread makers, or brewers). The hygienic standards for organic products under EU legislation are largely devised for large-scale economies, disregarding, and in some cases rendering almost impossible, the production of small-scale producers. It is for this reason that Campi Aperti's label is called "Genuino Clandestino" (*Genuine Clandestine*). Each certified product carries this label, jointly with the description 'Community fighting for food self-determination', a notion that could be interpreted as bridging the idea of food sovereignty, autonomy, and the commons.

Because farmers participating in Campi Aperti position themselves as a viable alternative to the agro-industries, the PGS is succinctly employed to overcome economic barriers (IFOAM, 2015). The high standards of the PGS combine the sustainable values of organic production applied at all stages of production with the focus on a farmer's control over seeds, plant, or manure inputs for cultivation, water and soil quality, as well as transport costs (all of them, food-producing commons). The introduction of external and oil-based energy inputs is banned, since agro-ecological farming and circular farming methods are meant to be the norm. In addition, dependent workers are included in the inspection process to avoid exploitation and abuse. In case a transformed product required an off-site processing site, for example a flour mill or an olive oil press, this processing site is also controlled for its sustainability standards, in particular regarding energy and waste.

In conclusion, the experience of Campi Aperti shows that solidarity across the food chain, the convergence of urban and rural resistance, and the collective construction of transformative strategy must be considered an integral part of the movement for agroecological cities and for the transition towards political and economic autonomy of the food chain. The fact that urban 'consumers' in Bologna are recognised (and named) by Campi Aperti as co-producers, and that many of them were crucial in setting up the markets in the city, may epitomise new productive solidarities or new ways of 'commoning' that open up alternative paths for agroecological urbanism.

Although the strategy, that could also be termed either alter- or counter-hegemonic, remains constant, the governance systems are continuously revised because of their dynamic engagement with self-managed economic experiments (i.e., the cost and price arrangements and the PGS). The attention paid to reflexivity attracts new producers and clients/co-producers in a continuous redefinition of the network and the interactions among people, spaces, and nature. Whilst the growth of the association is desirable to gain political bargaining power with the municipality and national authorities, new producers often do not necessarily grasp the full intention and meaning of political autonomy. In that sense, internal frictions and tensions surfaced when Campi Aperti was in negotiations with the state and local authorities, and questions have also been raised concerning the focus on the trading aspect of the food system rather than the 'commoning' of means of production (e.g., land), and the fact that (differently from Community Supported Agriculture schemes) risk is not shared between farmers and co-producers.

In this context, self-awareness, self-critique, dialogue, and transformation assume a central role in the ability to engage in 'commoning'. In light of this, Campi Aperti and its members have been exercising constant reflection on their political processes (i.e., reflexive governance after De Schutter and Lenoble, 2010) and are always looking for new ways to extend their complex horizontal governance systems in order to absorb the growing number of producers and co-producers without compromising their political and economic autonomies. In a European context where 70% of the population lives in cities, Campi Aperti and Genuino Clandestino offer a concrete example of solidarity, 'commoning', and self-governance as acts of resistance against the *status quo* that are the precondition to a just and agroecological transition.

Section VI – Conclusions

This chapter has exposed a series of, often not too evident, connections behind the food system:

- that between enclosure of the commons in the North and South and the growth of the imperial metropolitan city;
- that between the plantation system, the birth of capitalism, the objectification of nature, and the continuous disempowerment of women and indigenous groups who were relying the most on those commons;
- that between the reproduction of white male privilege and the impossibility of a just and agroecological transition.

Faced with these historical and contemporary interdependencies, the coalescence of different scholars' and activists' struggles for a gender-transformative (and anti-patriarchal), de-colonised and de-commodified food system emerges as a highly needed goal. The disentangling of those connections is nothing but a first step in the pursuit of a fair and sustainable food system. A second step lays in cultivating reflexivity and facilitating frequent interactions between urban political consumerism in the North (expressing mostly a reforming attitude) and the rural food sovereignty movements in the South (challenging the system from an oppositional/radical stance). This is in order to encourage the convergence of urban and agrarian food justice struggles, where urban green spaces become experimental grounds for the de-commodification of food and where urban food policies are constructed with and for the non-urban.

In this chapter, we explored alternatives to the commodity-based food system that considers nature as a mere object, labour as a cheap input that should be reduced to the minimum, and food as cheap energy for the body (Vivero-Pol et al., 2019). These paradigmatic changes towards food and the food system can inform alternatives such as agroecological urbanism. We understand this as a technical and political reaction to the food-disabling urban landscapes that have separated urban dwellers from the food they eat, from the nature that they exploit, and from the living conditions of the people that make food possible everywhere in the world (Tornaghi and Dehaene, 2020). The ideas of commons and commons-based food systems, 'commoning', and food as a commons, although conceptually different, can inform democratic, fair, and ecological food systems. This happens, as the three examples provided in this chapter discuss, by challenging the existing profit-maximising, individualistic, and exploitative organisation of food production, consumption and post-consumption that the industrial food system represents. Moreover, the examples presented here include not just different praxis but, more importantly, values and narratives of food that drastically differ from the mainstream description of food as a commodity. The common thread of those cases is a pursuit of autonomy, self-determination by means of direct involvement in food production, consumption or governance, and a revaluation of food as a multidimensional essential good that is historically defined, intersectional, and intrinsically ecological.

In addition, the three cases presented in this chapter pivot around urban-rural interconnections, cross-chains solidarity, giving visibility to food people and food spaces that are often forgotten, and strengthening intersectional and historical self-awareness. In that sense, they epitomise what Arturo Escobar called an 'autonomous design' that eschews commercial and modernising aims in favour of more collaborative and place-based approaches (Escobar, 2018). Such a design can be based on the radical interdependence of all beings, or mutual neediness, as posited by philosopher John O'Neill in a recent essay (O'Neill, 2019). This autonomous redesign can be done through the assumption of relevant knowledge (e.g., cuisine recipes, agrarian practices, and public research), nature (e.g., seeds, fish stocks, land, forests, and water), social relationships, and more (e.g., solidarity, equality, justice, conviviality, anti-patriarchy, and anti-colonialism) as commons or collective practices. Those knowledge, natural resources, and behaviours can be co-constructed, valued, and governed as commons by means of self-regulated arrangements. And that collective reconstruction can inform the transformative ideas and movements that aim to change the industrial food system, such as the right to food, food sovereignty, food justice, or food democracy.

While this chapter welcomes political approaches to the commons that see commons as key innovative and transformative tools to help food systems move beyond the state and the market (Bollier and Helfrich, 2015), more debate and research is needed on the possible tensions and synergies that could be found with the discourses on human rights, food sovereignty and agroecological urbanism. However, we should not spend too much time thinking without practising. The commons have traditionally been created by the instituting power of collectivities acting together under self-regulated rules (Dardot and Laval, 2015). Moreover, 'commoning' does require nothing more than imagining, performing, and experimenting together for a common purpose. The risk of excessive reflection is that, as in the past, it would give space for legalisation, institutionalisation, and co-optation by institutionalised powers (i.e., landlords, kings, or nation-states). This tension between who and how the commons can be created, re-created, or legitimately recognised is at the core of the political and moral debate, triggering discrepancies and conflicts that must be mapped, considered and addressed.

Notes

1 N.A.: Translated into English ("Eating together makes the exercise easier").
2 Although we emphasise here the role played by capitalist actors in the encroachment and commodification of natural resources, we do not forget or understate the importance of state appropriation of natural resources owned and governed by collectivities. Both state and private appropriation of people's commons are mutually reinforcing processes. For instance, in post-colonial Africa, the state is responsible for land grabbing and dispossession of customary land rights that were then transferred to international foreign corporations under long-term leasing agreements.
3 Throughout the 20th century, a few Western economists established a theoretical framework to classify all types of goods into four categories, based on rivalry and excludability. After that, food was considered a private good and therefore a perfect subject to be allocated exclusively through market mechanisms.
4 Actually, commoning food in urban settings, mostly done by consumers that largely purchase food and are therefore consumers, has different features from commoning in rural areas, where more than three-quarters of rural inhabitants produce food themselves, either for self-consumption or selling.
5 For more on the six dimensions of food, i.e., food as an essential good, natural resource, human right, cultural determinant, tradeable good, and commons, see Vivero-Pol (2017).
6 https://undocs.org/pdf?symbol=en/A/RES/73/165.
7 This section builds on a conference paper presented by Stefania Errico and Priscilla Claeys at the 2017 Utrecht conference of the International Association for the Study of the Commons (IASC) under the title: 'Human Rights and the Commons: Bridging Gaps and Exploring Complementary Approaches to the Governance of Land and Natural Resources'.

8 Many more papers on different issues can be found at: https://landworkersalliance.org.uk/publications/.
9 https://www.landjustice.uk.
10 The 2nd Nyéléni Europe Forum for Food Sovereignty took place from 26 to 30 October 2016 in Cluj-Napoca, Romania.
11 To know more on Sylvia Wynter's thoughts on decolonialism, see McKittrick (2015).
12 Pisa, B. Interview done on 27 February 2018 by Dagmar Diesner for research on her Phd 'Reframing food sovereignty in the context of Italy: Barriers and resilience strategies in the conceptualisation of commoning on the case study of Campi Aperti'.
13 In Campi Aperti, participants have decided to frame the urban consumers as 'co-producers' instead of mere consumers to emphasise the solidarity and connectivity between rural-based and urban-based components of the movement.

References

Akala (2018) *Natives: Race and Class in the Ruins of Empire*. London: Two Roads Books.

Anghie, A. and Chimni, B. S. (2003) Third world approaches to international law and individual responsibility in internal conflicts. *Chinese Journal of International Law*, 2(1), pp.77–103.

Arrighi, G. (1994) *The Long 20th Century. Money, Power, and the Origins of Our Times*. London: Verso.

Batliwala, S. (2010) *Feminist Leadership for Social Transformation. Clearing the Conceptual Cloud* [online]. Bangalore: Creating Resources for Empowerment in Action (CREA).

Bernstein, H. (2016) Agrarian political economy and modern world capitalism: the contributions of food regime analysis. *The Journal of Peasant Studies*, 43(3), pp.611–647.

Bollier, D. and Helfrich, S. (eds) (2015) *Patterns of Commoning*. Amherst: Commons Strategy Group and Off the Common Press.

Bollier, D. and Helfrich, S. (2019) *Free. Fair and Alive. The Insurgent Power of the Commons*. New Society Publishers. http://www.freefairandalive.org/read-it/#NotePublisher.

Borras, S. and Franco, J. (2012) *A 'Land Sovereignty' Alternative? Towards a People's Counter-Enclosure*. Amsterdam: TNI Agrarian Justice Programme.

Bradley, K. and Herrera, H. (2016) Decolonizing food justice: naming, resisting and researching colonizing forces in the movement. *Antipode*, 48(1), pp.97–114.

Butterly, D. and Fitzkpatrick, I. (eds) (2017) *A People's Food Policy. Transforming Our Food System*. www.peoplesfoodpolicy.org.

Campi Aperti (2014) *Carta dei Principi*. Available at http://www.campiaperti.org/chi-siamo/carta-dei-principi/ (acessed 20.01.2019).

Chakraborty, R. (2014) The Bengal famine: how the British engineered the worst genocide in human history for profit. YourStory Media Pvt Ltd, 15 August 2014. https://yourstory.com/2014/08/bengal-famine-genocide/ (accessed 5 February 2018).

Claeys, P. (2015) *Human Rights and the Food Sovereignty Movement: Reclaiming Control*. New York: Routledge.

Claeys, P. (2019) The rise of new rights for peasants. From reliance on NGO intermediaries to direct representation. *Transnational Legal Theory*, 9(3–4), pp.386–399.

Cohen, E., Reynolds, K. and Sanghvi, R. (2012) *Five Borough Farm: Seeding the Future-of Urban Agriculture in New York City*. New York City, NY: Design Trust for Public Space.

Committee on World Food Security (CFS) (2012) *Voluntary Guidelines on the Responsible Governance of Tenure of Land, Fisheries and Forests in the Context of National Food Security*. http://www.fao.org/cfs/home/activities/vggt/en/.

Counihan, C. (2016) Cultural heritage in food activism: local and global tensions, in Brulotte, R. L. and Di Giovine, M. A. (eds) *Edible Identities: Food as Cultural Heritage*. Farnham, UK and Burlington, VT: Routledge, pp.219–229 .

Dalla Costa, M. (2007) Food sovereignty, peasants and women. *The Commoner.* Available at https://docplayer.net/121404253-Food-sovereignty-peasants-and-women.html (accessed 10.03.2020).

Dardot, P., and Laval, C. (2015) *Commun: essai sur la révolution au XXIe siècle.* Paris: La Découverte.

De Angelis, M. (2017) *Omnia Sunt Communia. On the Commons and Transformation to Postcapitalism.* London: Zed Books.

De Schutter, O. and Lenoble, J. (eds) (2010) *Reflexive Governance: Redefining the Public Interest in a Pluralist World.* Oxford: Hart Publishing.

De Sousa Santos, B. (2014) *Epistemologies of the South: Justice against Epistemicide.* Abingdon: Routledge.

De Sousa Santos, B. (2018) *The End of the Cognitive Empire: The Coming of Age of Epistemologies of the South.* Durham, NC: Duke University Press.

Deh-Tor, C. M. (2017) From agriculture in the city to an agroecological urbanism: the transformative pathway of urban (political) agroecology. *Urban Agriculture Magazine*, 33, pp.8–10. https://www.ruaf.org/sites/default/files/RUAF-UAM%2033_WEB_8-10.pdf.

Errico, S. (2017) Control over natural resources and protection of the environment of indigenous territories, in Weller, M. and Hohmann, J. (eds) *The UN Declaration on the Rights of Indigenous Peoples.* A Commentary, Oxford: Oxford University Press, pp. 425–460.

Escobar, A. (2018) *Designs for the Pluriverse: Radical Interdependence, Autonomy, and the Making of Worlds.* Chapell Hill: Duke University Press.

Fanon, F. (1963) *Black Skin, White Mask.* Harmondsworth: Penguin.

Federici, S. (2004) *Caliban and the Witch. Women, the Body and Primitive Accumulation.* New York: Autonomedia.

Federici, S. (2019) *Re-enchanting the World: Feminism and the Politics of the Commons.* Oakland, CA: PM Press.

Ferrando, T. (2018) Financialisation of the transnational food chain: from threat to leverage point? *Transnational Legal Theory*, 9(3–4), pp.316–342.

Ferrando, T. and Vivero-Pol, J. L. (2017) Commons and commoning: a 'new' old narrative to enrich the food sovereignty and right to food claims. *Right to Food and Nutrition Watch*, 10, pp.50–63.

Friedmann, H. and McMichael, P. (1989) Agriculture and the state system: the rise and fall of national agricultures, 1870 to the present. *Sociologia Ruralis*, 29(2), pp.93–117.

Garnett, S. T., Burgess, N. D., Fa, J. E., Fernández-Llamazares, A., Molnár, Z., Robinson, C. J., Watson, J. E., Zander, K. K., Austin, B., Brondizio, E. S. and Collier, N. F. (2018) A spatial overview of the global importance of Indigenous lands for conservation. *Nature Sustainability*, 1(7), pp.369–374.

Hamilton, J. W. and Bankes, N. (2010) Different views of the cathedral: the literature on property law theory, in McHarg, A., Barton, B, Bradbrook, A. and Godden, L. (eds) *Property and the Law in Energy and Natural Resources.* New York: Oxford University Press.

Haraway, D. (2015) Anthropocene, capitalocene, plantationocene, chthulucene: making kin. *Environmental Humanities*, 6(1), pp.159–165.

Harper, A. B. (2016) Doing veganism differently: racialized trauma and the personal journey towards vegan healing, in Hayes-Conroy, A. (ed.) *Doing Nutrition Differently.* London: Routledge, pp.151–168.

IFOAM (2015) *Participatory Guarantee Systems (PGS)*. International Organisation for Organic Agriculture. Available at http://www.ifoam.bio/en/value-chain/participa tory-guarantee-systems-pgs (accessed 12 January 2019).

Kay, S. (2016) *Connecting Smallholders to Market: An Analytical Guide*. The Civil Society Mechanism (CSM), United Nations Committee on World Food Security.

Kothari, A., Salleh, A., Escobar, A., Demaria, F. and Acosta, A. (eds) (2019) *Pluriverse: A Post-Development Dictionary*. Delhi: Authors Up Front.

Light, A., Simpson, G. and Weaver, L. and Healey, P. G. (2009) Geezers, turbines, fantasy personas: making the everyday into the future. *Proceedings of the 7th Conference on Creativity & Cognition*, Berkeley, California, USA, October 26–30, pp.39–48.

Linebaugh, P. (2008) *The Magna Carta Manifesto. Liberties and Commons for All*. Berkely: University of California Press.

Mama, D and Anderson, C. (2016) *Decolonisation and Food Sovereignty in Europe, Thoughts from the Edges*. Available at http://www.peoplesknowledge.org/discussions-on-dec olonising-food-food-sovereignty-in-europe/

Mattei, U. (2013) Protecting the commons: water, culture, and nature. The commons movement in the Italian struggle against neoliberal governance. *South Atlantic Quarterly*, 112(2), pp.366–376.

Mattei, U. and Nader, L. (2008) *Plunder: When the Rule of Law is Illegal*. Oxford: Blackwell Publishing.

Maughan, C. and Ferrando, T. (2019) Land as a commons: examples from the UK and Italy, in Vivero-Pol, J. L., Ferrando, T., De Schutter, O. and Mattei, U. (eds) *Routledge Handbook of Food as a Commons*. Abingdon: Routledge, pp.329–341.

McKittrick, K. (ed.) (2015) *Sylvia Wynter: On Being Human as Praxis*. Chappell Hill: Duke University Press.

McMichael, P. (2005) Global development and the corporate food regime, in Buttel, F. H. and McMichael, P. (eds) *New Directions in the Sociology of Global Development Volume 11*. Bingley, Warwickshire: Emerald Group, pp.265–299.

Mignolo, W. D. (2011) *The Darker Side of Western Modernity: Global Futures, Decolonial Options*. Durham: Duke Press.

Moore, J. W. (2015) *Capitalism in the Web of Life: Ecology and the Accumulation of Capital*. London and New York: Verso Books.

Moyles, T. (2015) Permaculture or spermaculture? Confronting patriarchy in permaculture and alternative food movements. *Briarpatch*. Available at https://br iarpatchmagazine.com/articles/view/permaculture-or-spermaculture (last visited 18 November 2019).

NBFJA (2017) *NBFJA Statement: Land Justice is Essential to Food Justice*. National Black Food & Justice Alliance. Available at https://www.blackfoodjustice.org/nbfjablackfo odblog/2018/3/19/7fdjpr6yaeh8cebhc9pk4ktypeifn6 (last visited November 18, 2019).

O'Neill, J. (2019) Food, needs, and commons, in Vivero-Pol, J. L., Ferrando, T., De Schutter, O. and Mattei, U. (eds) *Routledge Handbook of Food as a Commons*. Abingdon: Routledge, pp.103–120.

Ostrom, E. (1990) *Governing the Commons. The Evolution of Institutions for Collective Action*. New York: Cambridge University Press.

Patel, R. and Moore, J. W. (2017) *A History of the World in Seven Cheap Things. A Guide to Capitalism, Nature, and the Future of the Planet*. Berkeley: University of California Press.

Pettenati, G., Toldo, A. and Ferrando, T. (2019) The food system as a commons, in Vivero-Pol, J. L., Ferrando, T., De Schutter, O. and Mattei, U. (eds) *Routledge Handbook of Food as a Commons*. Abingdon: Routledge, pp.42–56.

Rights and Resources Initiative (RRI) (2015) *Who Owns the World's Land? A Global Baseline of Formally Recognized Indigenous and Community Land Rights.* Washington, DC: RRI.

Rivera-Ferre, M., Constance, D. and Renard, M. (2014) Convergence and divergence in alternative agrifood movements, in Constance, D. H., Renard, M. C. and Rivera-Ferre, M. (eds) *Alternative Agrifood Movements: Patterns of Convergence and Divergence. Research in Rural Sociology and Development,* vol 21, pp.313–322.

Said, E. (1978) *Orientalism.* London: Routledge & Kegan Paul.

Sassen, S. (2014) *Expulsions: Brutality and Complexity in the Global Economy.* Cambridge: Harvard University Press.

Schiller, R. J. (2019) *Narrative Economics: How Stories Go Viral and Drive Major Economic Events.* Princeton, NJ: Princeton University Press.

Shawki, N. (2015) Transnationalism and diffusion: a study of the food sovereignty movements in the UK and Canada. *Globalizations,* 12(5), pp.758–773.

Sibley, D. (1995) *A Geography of Exclusion.* London: Routledge.

Slocum, R. (2007) Whiteness, space, and alternative food practice. *Geoforum,* 38(3), pp.520–533.

Szymanski, I. F. (2016) What is food? Networks and not commodities, in Rawlinson, M. and Ward, M. C. (eds) *The Routledge Handbook of Food Ethics.* Abingdon: Routledge, pp.7–15.

Thornberry, F. and Viljoen, F. (2009) *Overview Report of the Research Project by the International Labour Organization and the African Commission on Human and Peoples' Rights on the Constitutional and Legislative Protection of the Rights of Indigenous Peoples in 24 African Countries.* ILO: Geneva.

Tornaghi, C. (2017) Urban agriculture in the food-disabling city: (re) defining urban food justice, reimagining a politics of empowerment. *Antipode,* 49(3), pp.781–801.

Tornaghi, C. and Dehaene, M. (2020) The prefigurative power of urban political agroecology: rethinking the urbanisms of agroecological transitions for food system transformation. *Agroecology and Sustainable Food Systems,* 44 (5), pp. 594–610.

United Nations (2009) *General Comment No. 12* – Committee on Economic, Social and Cultural Rights, UN Doc. E/C.12/1999/5.

United Nations General Assembly (2018) *United Nations Declaration on the Rights of Peasants and Other People Working in Rural Areas.* 30 October 2018, UN Doc. A/C.3/73/L.30.

United Nations Human Rights Council (2012) *Final Study of the Human Rights Council Advisory Committee on the Advancement of the Rights of Peasants and Other People Working in Rural Areas.* 24 February, UN Doc. A/HRC/19/75.

Visser, O., Mamonova, N., Spoor, M. and Nikulin, A. (2015) 'Quiet food sovereignty' as food sovereignty without a movement? Insights from post-socialist Russia. *Globalizations,* 12(4), pp.513–528.

Vivero-Pol, J. L. (2017) Food as commons or commodity? Exploring the links between normative valuations and agency in food transition. *Sustainability,* 9(3), p.442.

Vivero-Pol, J. L. (2019) The idea of food as a commons: multiple understandings for multiple dimensions of food, in Vivero-Pol, J. L., Ferrando, T., De Schutter, O. and Mattei, U. (eds) *Routledge Handbook of Food as a Commons.* Abingdon: Routledge, pp.25–41.

Vivero-Pol, J. L. Ferrando, T., De Schutter, O. and Mattei, U. (2019) *Routledge Handbook of Food as a Commons.* Abingdon: Routledge.

Wall, D. (2014) *The Commons in History. Culture, Conflict, and Ecology.* Cambridge, MA: MIT Press.

Wallerstein, I. (1974) *The Modern World-System I: Capitalist Agriculture and the Origins of the European World-Economy in the Sixteenth Century.* New York: Academic Press.

Wynter, S. (2003) Unsettling the coloniality of being/power/truth/freedom: towards the human, after man, its overrepresentation—an argument. *The New Centennial Review,* 3(3), pp.257–337.

4

URBAN AGRARIAN ALLIANCE BUILDING IN PERI-URBAN ROME

The pivotal role of land access in food system reconfiguration

Luca Colombo, Stefano Grando and Giacomo Lepri

1. Land ahoy! Farmland in sight in peri-urban areas, but not necessarily within reach

Farmland in Europe is subject to competition on its use, control and function, on both its tangible and intangible determinants. When assessing land tenure and land access in Europe, in addition to land availability, other factors should also be considered; these include the development in land concentration, the ongoing process of structural change of EU agriculture, the role that policies have on this process, land and 'green' grabbing, as well as land acquisitions in the name of climate change mitigation strategies, such as the installation of solar panels (Borras et al., 2016; van der Ploeg et al., 2015; Kay et al., 2015; Fairhead et al., 2012). These processes may have relevant effects on the availability and use of agricultural land and, indirectly, on food and nutrition security (FNS); this is due to changes in yields, employment, food availability and the role of family farming.

As a result of farmers' retirement, change of jobs, or farm bankruptcy, an ongoing trend of land concentration due to acquisitions within the primary sector is occurring across Europe (van der Ploeg et al., 2015). This is leading to takeovers by larger neighbouring and non-neighbouring farms; this is in parallel with speculative and investment activities by non-agrarian sectors such as those performed by building and energy companies. At the same time, loss of farmland due to urbanisation, farm abandonment and rural migration, or declining soil fertility has been compensated for by new land being placed under cultivation. However, this has not been without environmental costs, as shown by forests being replaced by intensive cultivation of high-value crops (e.g., vine or apple plantations in Northern Italy). In these very situations, the hunt for land is both an effect and a cause of rise in land values (up to €200,000/ha for a vineyard or orchard in Trentino; Gioia and Mari, 2012), making access to land for the non-bankable farming population more difficult, particularly for young farmers.

In a country like Italy, trends in land values are conditioned by the land's potential: on the one hand, the land market is stagnant in inner and so-called marginal areas with low and stable land prices; on the other hand, there is an upward trend in land prices in fertile plains and in economically dynamic territories, also considering alternative land uses (Mela et al., 2012). In particular, areas suitable for good profitability within and outside the primary sector, such as those located next to cities, continue to attract investments, often determining the scale and relevance of appetite for land and related conflicts.

In the context of farmland access, peri-urban areas show a specific peculiarity in consideration of the producers' difficulties in acquiring land and securing its tenure, despite the growing attention that this issue receives in the local urban policy agenda, particularly in Southern Europe (Cerrada-Serra et al., 2018). These hitches may, in fact, be more severe in urban and peri-urban areas where the food-feed-fibre-energy production competes with urban settlements, infrastructure development and recreational drivers. Therefore, the growing scarcity of fertile land and the urban sprawl, in particular when associated with land fragmentation and soil degradation driven by illegal housing and real estate speculation, are currently escalating citizens' interest in the quality of their living environments. A mounting awareness about the importance of saving/reintroducing high-quality green areas (Žlender and Ward Thompson, 2017) tends to be accompanied by a request for policies mitigating soil consumption and promoting regenerative forms of agriculture on fringe land. Similarly, public administrations are also demanding a more energetic territorial presidium (Salvati, 2016). Moreover, as argued by Cerrada-Serra et al. (2018), cities in Mediterranean Europe tend to maintain strong relationships with their surroundings, driving a renewal of the links between urban development and agriculture, which are also a result of the recent economic crisis and emerging food-related social movements. As large parts of the Southern European population still suffer from various forms of 'food poverty' and continue to experience social exclusion, often exacerbated by the food system characters themselves (e.g., low-paying jobs for foodworkers and poor access to land) (IPES-Food, 2018), peri-urban agriculture and related access to land are seen as potential levers for poverty and unemployment alleviation.

Often referred to as the 'young farmer problem' (Hamilton et al., 2015; Zagata and Sutherland, 2015), unemployment (coupled with the ageing population of Italian and EU farmers) is a mounting political concern triggering policy discussions on access to land for a new farming generation. Rhetoric on young farmers is often associated with their predisposition to innovation, to market connection and to technology uptake/development (Leonard et al., 2017; Zagata and Sutherland, 2015). As such, they represent a factor of modernisation. However, the 'young promise' in agriculture is hindered by a 'critically low' availability of farms (Leonard et al., 2017). This is even more critical in peri-urban areas due to sharp competition by the building sector, and by an almost insurmountable problem of access to credit when lacking bankable assets to guarantee loans, as generally occurs with young unemployed people.

At the same time, peri-urban agriculture is receiving growing attention in Italy, as elsewhere in Europe. New farming initiatives proliferate, combining economic returns with socio-environmental ambitions, or proximity markets with innovative marketing formulas often designed by young entrepreneurs. This is giving rise to an increasing demand for land by both farmers and 'neo-rural' people.

Stimulated by concerns for the welfare of the younger generation of workers, growing civic concerns about the strategic role of fertile land and worries about its consumption and misuse have, in Italy, recently been tentatively addressed by some policy measures meant to support innovative land tenure initiatives that respond to both land use and employment stakes. Moreover, this is paralleled by civil society organisations and local authorities that, in an extended governance framework, set the stage for healthy local foods and sustainable agricultural practices to nurture local communities and economies (Brunori et al., 2016).

To identify transition pathways to sustainable and equitable food and nutrition futures for Europe, in the context of the EU-funded TRANSMANGO project, a case study was carried out in Rome, Italy, to examine how the allocation of public land to young farmers may trigger quality food provision, employment opportunities, social and environmental benefits and the reconfiguration of peri-urban areas and food systems. Rome, in fact, recently witnessed grassroots initiatives to grant would-be farmers access to land through the distribution of public land. In particular, in 2014, a network set up by a group of young (aspirational) farmers and various organisations and potential beneficiaries launched a mobilisation that contributed to an important achievement. This was the publication of two tenders, overall allocating ten farmland units held in public hands, as a pilot to mobilise both public and private vacant land. The case study aimed to examine how this allocation of public land actually responded to the activists' goals and was capable of stimulating the reorganisation of the welfare and local food systems.

After the Italian agricultural reform of the 1950s (2,844 farm units allocated in the Lazio Region out of 68,000 in Italy, overall redistributing 760,000 hectares; Ghini, 2012) and the land occupations of the late 1970s (mostly led by groups of young unemployed and politicised people claiming control over public lands), Rome's recent allotment assignments of peri-urban areas reflected a relaunch of policy initiatives on land tenure triggered by struggles for social redistribution of economic means.

The case study findings indicate that Rome's mobilisation on land access epitomises innovative reconfigurations of modern stakes in western societies, such as the search for a job based on ethical entrepreneurship, a socially fair utilisation of public land to develop multifunctional farming (Di Iacovo, 2011, Zasada, 2011), the delivery of fresh and seasonal local food at reasonable prices and the promotion of more advanced food and nutrition awareness.

These trajectories can be seen as niche initiatives, developed within nested markets that are, to a certain extent, detached from industrial food price competition (van der Ploeg et al., 2012). However, these niches can also be regarded as vanguards of a new way of approaching food production and consumption that may, in the future, develop towards a more diversified and resilient urban food system, or even towards more radical social change in the relationship around communities, land, territories and food.

On the basis of the reflections raised by this experience, here we discuss how the question of democratic land control is entwined with the broader idea of an alternative food system (Franco et al., 2015), and how the land access struggle converges with the development of new organic farms, analysing the initiatives' transformational potential.

2. Participatory observations to assemble the evidence mosaic

The case study investigation used various methodological tools. The approach included participatory observations, desk analysis and interviews, enriched by a scenario workshop articulated into two separate meetings. The familiarity the researchers had with the activists, due to previous personal engagement in food sovereignty and short food circuits-related initiatives, made the whole process doable and meaningful. Trust and acquaintance among those involved created an enabling environment for the study and allowed a free flow of reflections. However, researchers decided to maintain a distinctive role and display the methodology commonly agreed within the European project.

The desk based analysis framed the information acquired from primary sources. The study scrutinised policy documents, mass media articles and online news pertinent to land access and urban agriculture in Rome, including internal social organisation documents and press releases. This work led to the assessment of the wider territorial and sociocultural context in which the initiatives for land access developed, the main phases of the mobilisations carried out by the activists, and their achievements. Scientific literature was also explored to locate emerging reflections in the wider academic debate.

To offer participants room for visioning and to grant researchers insights from a heterogeneous collective, a two-day participatory scenario workshop (16th February 2016 and 1st April 2016) was organised with various actors (farmers and would-be farmers, local administrators, environmentalists, civil society activists and organic associations' managers) and researchers involved mobilisation in recent years. The workshop played a central role in the case study development as it provided space for open debate and suggestions for individuals and organisations involved and/or interested in the initiatives. In particular, the workshop offered stakeholders the opportunity to discuss themes detached from the everyday issues and controversies, thus focusing on forthcoming opportunities and threats and envisaging future actions to be undertaken through cooperation among actors.

To discuss findings and interim outcomes, further to acquiring an understanding of the different positions, objectives and visions, interviews with key activists enriched the collected information. Particular attention was paid to good practices, critical and emerging issues around land access and land use, ideological and pragmatic considerations, hampering and enabling factors, further opportunities and possible threats for the development of land access initiatives.

3. From mobilisation to assignments, from common visions to fragmented endeavours

The renewed interest in access to agricultural land in Rome

In recent years, a range of innovative models of short chain delivery of food have emerged in Rome. The presence of traditional forms of family farming, which have survived throughout the city's modernisation, has developed into a landscape of grassroots initiatives around food and agriculture; these have grown increasingly animated in the last couple of decades (Fonte, 2013; Grando et al., 2017). This traditional farming is promoted by both producers and consumers, including through joint efforts. These re-localisation and short circuit initiatives, which have received mounting interest for their multidimensional benefits (Zasada, 2011; Mok et al., 2014), are occasionally supported by public bodies and are still spreading at both city and regional levels. The rationale for most of these initiatives is the direct relationship between organised consumers and individuals, or aggregated farmers producing in areas near Rome through organic farming or similar methods.

These experiences contribute to the revitalisation of the agricultural sector in the metropolitan area of Rome and, coupled with youth unemployment, trigger increasing awareness of agriculture and alternative ways of farming. This is anecdotally evidenced by the proliferation of urban gardens and the success of well-attended training initiatives on organic agriculture organised by organic farming and environmental associations.

More strategically, this renewed interest in peri-urban agriculture is also witnessed by aggregation processes of groups of young individuals mobilising for access to land vis-à-vis the public administration. The *Coordinamento Romano per l'accesso alla terra* (CRAT, Roman coordination for access to land), a network set up by a group of young would-be farmers and various organisations and potential beneficiaries, was particularly active on land access and use issues. Starting from the spring of 2011, groups of young farmers, aspiring farmers, agricultural cooperatives born from the squatting of public areas during the 1970s, NGOs, trades unions and environmental associations joined together in CRAT. The coordination united formal organisations, informal groups and individuals with the stated goals of resuming social and political initiatives to democratise land access and use, and to mobilise for a renewed public policy agenda around farming land and local food systems. The economic and employment crises, the legislative discussion on long-term lease or acquisition of land for young farmers and a mounting interest in land use and soil depletion provided a favourable momentum for the mobilisation. Symbolic initiatives (ranging from flash mobs to a large mural in Rome, symbolic occupations of abandoned areas, training seminars and active participation in academic and institutional conferences and public debates) attracted political attention: on those occasions, CRAT defended the principles of granting access to land for disadvantaged individuals (unemployed people and landless farmers), who were otherwise unable to access it, and the importance of multifunctionality in agriculture, particularly in peri-urban areas.

In fact, CRAT strategically chose to take the legal route; this was in order not to appear troublesome to local communities and the larger public, ultimately being both potential allies and recipients of the benefits arising from young farmers accessing the land. For this reason, instead of land squatting, CRAT opted to call on local administrations to ease access for all public lands, and for all young people to secure the holdings through ten-year contracts that could shelter new entrants in their agricultural business start-up. The hope was that such a pact would guarantee collaboration between the private and public sector, facilitating access to funds and providing training; this invigorated the initiative. The reasons and aims of the mobilisation were expressed in a manifesto that itself became a catalyst and a token of the movement (CRAT, 2011a). The document expressed the need for clear and transparent procedures for public land allocation to young farmers, with the aim of recovering abandoned plots of land while establishing a new alliance between farmers and consumers towards a new rural-urban relationship.

In a public speech made in 2011 by one of the leaders of the mobilisation, the aims of the CRAT initiative were summarised by recalling and expanding some of the manifesto key statements through the description of four wide-ranging needs: the need for employment ('satisfying jobs that people are free to choose and that would make them feel rewarded'); the need for space ('to counter the shortage of public spaces and to connect city dwellers with the farming community'); the need for a liveable city ('a city of efficiency, able to produce the goods necessary for its survival, self-sufficient and green'); and the need to use abandoned public spaces ('because many of the forgotten green areas in the city represent environmental and social values, and they must be entrusted to those who want to take care of and valorise them').

Moreover, the mobilisation acknowledged that the urban poor continue to lack access to nutritious food, either because of lack of physical access to healthy food choices in poor neighbourhoods or because of lack of income, as occurring in *obesogenic food environments* (Swinburn et al., 1999) like the city outskirts. This also recalls the pedagogic mission the land access movement imagined when advocating for organic production for local dwellers.

This initiative finally succeeded in gaining institutional attention. Farms held in public hands and suitable for farming became the object of negotiations with municipal and regional authorities; these negotiations aimed to make the land available for young farmers with advantageous terms. Farmland units were also mapped and pinpointed to operationalise the assignments. In 2014, at the peak of the mobilisation, a tender was published by the Rome Municipal authority allocating for long-term leases four lots of land accounting for 97 hectares in total. Coupled with the tender issued by the Regional Administration, 10 areas, totalling about 400 hectares, have been made available: a small, but symbolically important number of farm units.

Some of the identified allotments resulted from a peculiar legal framework, the Municipal Master plan adopted in 2008, which allows the city administration to own or control an increased acreage of land that could be made available. This is a consequence of the implementation of the 'compensation planning system', which brings new environmentally sensitive areas under the management of the Municipality. According to the compensation system, the construction of new buildings can be prohibited by the Municipality in areas already owned or acquired by a building company. When this happens, the company is then allowed to build elsewhere, but only if the restricted area is handed over to the Municipality. These swaps imply the gratuitous land transfer to the Municipality and integrative urban qualification works. About 700 hectares of potentially available public land suitable for agriculture in the Rome Municipality resulted from such compensations.

Access to land as a 'just' prerequisite

The flagship case of Borghetto San Carlo, as well as other pilot cases of public land grants in Rome, is a good example of the above-mentioned terms. On the ancient Roman Via Cassia, surrounded by populated neighbourhoods between the city and the large archaeological Parco di Veio, the 22-hectare farm was allocated to the Coraggio Co-operative through a municipal tender in 2014. The allotment was based on a business plan that included organic farming, services related to home-grown vegetables, such as on-farm catering, urban gardens for the neighbourhood, educational and training activities, picnic areas, cycle paths, kindergartens, social integration opportunities, heritage orchards, ecological infrastructures and hospitality for the Via Francigena pilgrims.

The land was previously owned by the Impreme building company; this company signed a compensation deal with the Rome Municipality that also contemplated the refurbishment of the 1,400 square metre abandoned historic house part of the assignment (for an estimated spend of €2.5 million), based on the above-mentioned compensation criteria. The deadline for the house refurbishment, initially set for 2013, was delayed to 2016. However, renovation works only started in early 2019 after the scandal was covered by media and due to intense lobbying efforts triggered by the Coraggio Co-operative. Should it be completed, this would represent the first fully accomplished compensation in Rome. In the meantime, the cooperative remains without suitable shelters for the tractors and other vehicles, and with no or limited capacity for catering activities or other touristic opportunities (agritourism, hospitality) as planned at the onset of the project. The cooperative income generation has, therefore, been partially compromised, threatening the members' livelihood and the group's solidity itself.

In a situation where no investment or maintenance was carried out by the private owners, and despite the degradation of both the farming land and the existing used-to-be-rural premises, the co-op managed to grow a number of diversified crops (durum wheat, sorghum, vegetables, pulses, potatoes), reclaim 1 hectare of olive trees, place beehives and plant a heritage orchard (6 species, 48 varieties), based on regenerative organic principles. A wooden structure, made entirely of reclaimed wood, now hosts the 'Scuola diffusa della terra Emilio Sereni' training courses for would-be farmers, currently in its third edition.

The Coraggio Co-operative offers an additional example of how land assignments are not *per se* sufficient conditions for the sustainable and firm establishment of a new entrants' farm. The Rural Development Plan, drawn up by the Lazio Region and based on the EU Rural Development Policy (through Measure 6.1), theoretically provides €70,000 of unrestricted funds for the 'First Settlement of Young Farmers', inspired by the need for generational renewal. Access to these initial funds was impossible for both the Coraggio Co-operative and other young farmers awarded by the public lands assignments as they were not able to benefit from this public financial contribution. This was due to the specific regional requirements that requested a VAT code less than 12-months old and an entry income equivalent to €15,000 p.a. (equal to 3 hectares of vegetable garden with greenhouses, all perfectly running). In most cases, a VAT number was already activated in the past by the assignees making access to these funds inapplicable, while the €15,000 entry income represented a problematic threshold for most assignees. In consideration of the criticisms raised on such paradoxical parameters initially adopted by the Lazio Region and due to the specific observations by the European Commission, these criteria were revised and loosened, subsequently making the RDP measure more accessible and inclusive.

As the exemplificatory case of the Coraggio Co-operative shows, land assignments should be complemented by other enabling conditions, such as training or facilitated access to funding or administrative simplifications. The EU Common Agricultural Policy 'new entrants' provisions for young farmers is an example of possible enabling policy, as they make available grants or tax reliefs to eligible recipients; these are based on the idea that young farmers and/or new entrants represent an opportunity to steer agriculture towards lower input-dependency and more knowledge-intensive farming systems. However, this shift remains dependent on the availability of enabling factors, such as existing infrastructures (not least high-speed Internet, technical training and financial credit), or the removal of disabling administrative barriers. The experiences of Borghetto San

Carlo and other assignees reveal how accompanying measures may remain inaccessible for new entrants and merely facilitate succession within existing farming families. Consequently, the lack of financial support for the land assignees not only hindered their on-farm viability, but also constrained a firmer establishment of innovative multifunctional actors on the local food market.

In this respect, it may be useful to recall what has been negotiated at international level on land issues, since public land allotments are part of the wider context of land policies, guidelines and legislations. The Committee on World Food Security's Voluntary Guidelines on the Responsible Governance of Tenure of Land, Fisheries and Forests (CFS-VGGT), negotiated during more than two years of tense discussions and adopted in May 2012, encouraged states to provide prompt, accessible and non-discriminatory services to protect tenure rights, to promote and facilitate the enjoyment of those rights and to resolve disputes (FAO, 2012). Granting land tenure, particularly to smallholders, is an important but not sufficient feature. The need for integrative supportive conditions is similarly heralded by the People's Manual, drafted by international social movements to facilitate interpretation and use of the VGGT (IPC, 2016). Quite interestingly, the CRAT activists evoked the VGGT during their mobilisation to further legitimise and contextualise their requests.

Turning policies and mobilisations into administrative acts

As argued above, access to land is not *per se* sufficient if supportive complementary measures are lacking. Moreover, at the local administration level, even a set of comprehensive policy acts may still be inadequate if administrative procedures fail to correctly and timely implement decisions. In the analysed case, as stressed by participants during scenario workshops, a scattered distribution of competencies among different administrative levels, and an unreliable and inconsistent political will, played a crucial role in preventing the full realisation of the assignment experiences in their actual multifunctional economic activity. Moreover, these experiences are exposed to changes in political settings at each election, possibly giving rise to different priorities and distinct reference constituencies for the incoming administration.

These uncertainties were recently reflected in activists' increasing reluctance, as we will show below, to maintain a direct dialogue with institutions like the Rome Municipality or the Lazio Region, considered as 'administrative monsters' with their bureaucratic labyrinths and their political instability.

This somewhat frustrating acknowledgement represents a discontinuity if compared to the first years of mobilisation: during the CRAT campaigning and lobby activity in 2013–2014, negotiations between social actors and local authorities were based on principles of autonomy, respect and responsibility, while ensuring a certain degree of collaborative relationship in writing administrative measures. This brought a consensus on the tender approach and some of its details. For a relatively short time, the bureaucratic machine effectively flanked and enacted a policy negotiation between social groups and local authorities. Both the process and results were, to some extent, welcomed by the access-to-land movement, which nevertheless underlined that it only represented a step in a longer path. The role that some activists played in the policy drafting process likely contributed to a positive view of the achievements.

Subsequently, due to the lack of continuity and support in relation to these experiences by the Municipal and Regional Administration, cautious satisfaction turned into pragmatic disillusion and strategic reorientation. Moreover, despite converging in analysis, the activists' agendas and priorities varied in consideration of their different objectives, with assignees primarily focused on their productive endeavour, including showcasing its emblematic capacity to demonstrate the assignments' sociopolitical validity.

The CRAT mobilisation therefore discontinued and changed nature, with some prioritising farm management for its economic and symbolic value, and others redirecting attention to other local political disputes. This also happened in consideration of the almost concomitant political collapse of the ruling administration. These different perspectives and pragmatic visions were clear during the scenario workshops and the interviews.

Land assignments and organic farming

The Rome Municipality tender was grounded on two basic axes: (i) applicants had to be either professional farmers or young individuals below 40 years, or cooperatives thereof, (ii) applicants had to present a technical proposal (business and development plan). Both these requirements were consistent with the political platform developed by the activists engaged in the mobilisation, who stressed that it was important that the assignments had to fulfil social and environmental goals, based on overarching multifunctionality principles. Several criteria had to be met by the contenders, and their bids were evaluated against seven parameters ranging from the applicant's agricultural competence to the robustness of their proposal to the proposed use of renewable energy. The seven conditions also included the adoption of organic farming methods; this parameter was one of the most relevant, scoring 15/100 points. The selected proposals were finally chosen out of a group of 104, of which 80% were submitted by young farmers and 34% by women: all assignees had presented business plans grounded on organic farming.

The results were consistent with the CRAT strategic orientation of having access to land as a crucial pillar in the achievement of food sovereignty. The food sovereignty perspective was explicitly argued as a cultural horizon framing the mobilisation, meaning, in this context, the establishment of socially driven procedures for land access, and of alternative agroecological food chains in the control of smallholders. In this respect, and as a key method of cultivation, organic or agroecological farming was intended as both a lever of political coherence and a legitimisation tool vis-à-vis the neighbouring citizens and the town at large. The attention paid to organic farming in both the tenders and the applications made by would-be farmers means not only the adoption of sustainable and healthy agricultural practices, but also the possibility of generating more remunerative produce and higher value-added food chains, to promote food values to consumers and to create networks of ecologically and socially committed actors: in essence, a new and alternative development model.

4. New peri-urban farms as living social labs

The renewed social and political interest in peri-urban agriculture, the search for new livelihood options in the context of economic and employment crises and the opportunities arising from proximity markets have recently given rise to critical mobilisations on land access and tenure in Southern Europe, led by newcomers and/or by existing farmers (Cerrada-Serra et al., 2018).

Other than a mass struggle for land by the rural peasantry, as happened in Italy in the 1950s or 1970s, the Rome land access case illustrates the emerging interest in developing green/social belts around cities, preserving areas for agricultural production for social integration and local food circuits that link peri-urban areas with urban dwellers and relevant markets.

In such fringe territories, further to retaining a just share of the food sale value, access to land plays a pivotal role in sustainably building farmers' livelihoods as ownership or control of the land represents a crucial enabling factor, possibly greater than for other means of production. In this respect, long-term land leases to new farmers and pilot farm projects may act as modern symbolic bridges among territories (farmland, buildings, road networks and ecological zones), people (producers, consumers, citizens and public administrators; with neo-rural farmers being a synthesis themselves) and functions (food production and provision, natural resource management, education and recreation).

In this rearticulated context, legitimacy and neighbourhood acceptance of the assignees play a crucial role in the establishment of new farms. The combination of organic farming with multifunctional and recreation activities, in addition to present viable business opportunities, provides a base for a social pact with local inhabitants and consumers, with whom many assignees aim to create networks of ecologically and socially committed actors.

Despite their limited scale, these experiences seem to respond to societal concerns: given the social and ecological commitment of the incoming farmers, the newly established farms are seen as environmental sustainability presidia for the surrounding territory (whose ecological conditions are also crucial for the development of individual farms). This is in addition to partners, as shown by neighbouring residents benefitting from small plots made available by the Coraggio Co-operative for urban gardening. Indeed, the convergence of different interests enabled the coalition to grow with non-farming constituencies (neighbouring inhabitants, pupils' families and local administrators), going beyond theoretical and ideological visions. As stated by the activists during the case study scenario workshop and interviews, re-establishing the social function of the land, addressing those citizens who have pragmatic approaches to their daily matters, might even reduce the potential for conflict between different visions on land uses (farming, housing and public gardens). Indirectly, therefore, the alternative local food systems embedded in these allotments show a strong potential for bridging interests and constituencies, and represent an opportunity for a reconsidered role of farming land, too, particularly in peri-urban areas.

This is witnessed by the range of activities being developed by the Coraggio Co-operative (urban gardens, training school on agroecological farming, picnic areas, catering, landscape and ecosystem service maintenance). It is also witnessed by the experience of other well-established agricultural cooperatives, such as Cobragor and Agricoltura Nuova, active since the late 1970s and a cradle for many current young farmers and activists. Beyond food production, a deeper transformative activity is represented by constant awareness and education. These are not only based on the organisation of seminars and other public events, but also on the presence of multifunctional farms that offer citizens an opportunity to spend some time in a still rural and agricultural environment, and to familiarise themselves with organic and agroecological principles.

This convergence of interests, values and opportunities was one of the points of departure and strength of the land access mobilisation, which represents an interesting example of social engagement and civic participation that, to a certain extent, proved successful in Rome and potentially replicable in other contexts. A transformational potential is clearly visible if we consider the two parallel processes of scaling-out (replication of the initiatives, both in the Roman context and elsewhere) and scaling-deep (influence of these initiatives on attitudes and behaviours of the political and business spheres: a sort of 'contamination of the mainstream') (Moore et al., 2015). Differently, as clearly stated by the access-to land campaigners, scaling-up, or the dimensional growth of newly established farms, is not considered an option by the activists who aim for a future with a greater number of small farms rather than fewer larger ones.

Access to land, short value chains and agroecology, a trinity for environmental and food justice movements advocating for food sovereignty, would then result in a rural-urban blend of territories, social groups, farming techniques, use of natural resources, know-how and cultures. Similarly, this trinity shows the potential to mend producer-consumer relationships in a larger citizen perspective. A soft physical transition between built spaces and farming areas may in fact go in parallel with the food system temporal transition towards an ecologically literate re-localisation of agri-food practices. As mentioned, organic and agroecological farming plays a central role in this paradigm change: they are placed at the core of the technical motivations would-be farmers envision for their farm management and are seen by both activists and administrations as a way to gain citizens' and market legitimacy. Moreover, they are considered an essential tool to deliver ecological services in areas exposed to various types of environmental aggressions.

5. Conclusions

Assessing the transformational potential of a group of initiatives also entails understanding for how long these enterprises can last, and overcoming challenges that could possibly develop. In this regard, these access-to-land initiatives were debated in terms of bunkers/outposts, with the functions of environmental presidia and social laboratory. These outposts can have both a defensive role (niches where novelties can develop that are partially protected from powerful mainstream forces) and a proactive transformative role when they establish a sort of neuronal network, connecting to each other and with other similar initiatives.

For these niches to function and expand, various enabling conditions are needed, such as an effective policy interlocution with supportive local and regional administration and a political valorisation of such living social labs. More generally, the presence of a conducive socio-institutional environment, the creation of synergies with the most sensitive mainstream players, and a greater awareness among the general public of consumers/citizens about the potential of these initiatives may act as crucial enablers. These factors can support a wider transformative process centred on small-scale farming and short food chains, but potentially capable of triggering a more ambitious and substantial change in the development model, and in political approaches.

In this respect, the food and farming ambit provides a good potential for a wider societal transformation; this is particularly true in Italy, where it not only represents an important economic sector, but also acts as a vehicle for broader arenas of sociocultural change. For this reason, interviewed activists deemed it necessary to upscale the ambitions and broaden areas of consensus through far-reaching political proposals, including food sovereignty, social housing, smallholders' access to markets and other aspects not only framed within purely land issues, while at the same time granting sustainability to ongoing processes.

The issue for these niche initiatives is, in fact, how to firmly embed their practices into a legitimising sociopolitical framework that may contribute towards resisting vis-à-vis unfavourable conditions such as erratic political attention and support, competition on land use and on market access vis-à-vis more powerful competitors, lack of financial or administrative assistance, poor services and infrastructures. These assignments clearly provide critical alternative models of land management and access, and they can be expanded further if corresponding social and/or political support is granted. Networking, knowledge and training are also crucial leverage points that help to pool *savoir-faire* and best practices on land, including fostering agroecological transition and generational renewal.

As indicated by scenario workshop participants, both governance and technical networks are of utmost importance. Governance ultimately implies politics, and both may be influenced by activism and mobilisations; dialogue and lobby efforts are similarly considered relevant and necessary to achieve results. The Rome mobilisation on land access, in this respect, does not seem so frustrating, and it provides useful insights into how to achieve more. It also shows that networks may be far-reaching, particularly if based on solidarity and sharing principles on concrete aspects, but their fragility and dynamic development have to be taken into account.

The misalignment in tactics and aims shown by the activists after the assignments highlights how conflicts may be stratified with disagreements at one level and re-composition at another level. The social use of land remained a common focus, but with important differences among the activists' groups. For the assignees, their practical work (as farmland reclamation, restoration of the existing infrastructures, or even a tractor reparation) was part of a pedagogic activity and political struggle ('we use public land and produce public goods, showing that it can be remunerative while keeping those lands under the public domain'[1]). For the other group of activists, the social use of land had to materialise through permanent political action and confrontation with local authorities, for instance on public farms having a symbolic status, granting their better use and functioning. This disagreement is certainly influenced by subjective conditions (the fact of having an allotment assigned or not), but it seems also to reflect a wider and more theoretical divergence of visions.

Based on these outcomes, it should be questioned if a persistent and united mobilisation would have forced public institutions to hold a focus on new allotments of still available public lands, despite changes in the political composition of the ruling administration, or if, on a more pragmatic and cynical basis, CRAT was able to value the existing momentum to achieve a concrete result. For some interviewed actors, CRAT was seen as an *ad hoc* coalition aiming to bid for published tenders and the subsequent allotments. However, it is regretted that no substantial follow up or relay to other would-be farmer groups could be nurtured by the activists, even if some signs in this direction are in sight.

Acknowledgement

This chapter has been elaborated in the framework of the TRANSMANGO project (EU 7 FP) Assessment of the impact of drivers of change on Europe's food and nutrition security (Grant Agreement No: 613532)

Note

1. Case-study activist interview.

References

Borras, S. M., Jr., Seufert, P., Backes, S., Fyfe, D., Herre, R., Michele, L. and Mills, E. (2016) *Land Grabbing and Human Rights: The Involvement of European Corporate and Financial Entities in Land Grabbing Outside the European Union*. Brussels: European Union, Policy Department, Directorate-General for External Policies. doi:10.2861/26.

Brunori, G., Galli, F., Barjolle, D., van Broekhuizen, R., Colombo, L., Giampietro, M., Kirwan, J., Lang, T., Maye, D., de Roest, K., Rougoor, C., Schwarz, J., Mathijs, E., Schmitt, E., Smith, J., Stojanovic, Z., Tisenkopfs, T. and Touzard, J. M. (2016) Are Local Food Chains More Sustainable than Global Food Chains? Considerations for Assessment. *Sustainability*, 8(5), p.449. doi:10.3390/su8050449.

Cerrada-Serra, P., Colombo, L., Ortiz-Miranda, D. and Grando, S. (2018) Access to Agricultural Land in Peri-Urban Spaces: Institutional Frameworks, Land Use and Social Movements in Rome and Valencia. *Food Security*, 10(6), pp.1325–1336.

CRAT (2011). Vertenza per la salvaguardia dell'Agro Romano - Terre pubbliche ai nuovi agricoltori. Available at: http://accessoallaterra.blogspot.com/p/vertenza-per-la-salvaguardia-dellagro.html.

Di Iacovo, F. (2011) Le opportunità dell'agricoltura multifunzionale e sociale per l'agro-romano. In *Proceedings of the Workshop "L'Agro Romano tra tutela e sviluppo"*, Rome 15 June 2011.

Fairhead, J., Leach, M. and Scoones, I. (2012) Green Grabbing: A New Appropriation of Nature? *Journal of Peasant Studies*, 32(9), pp.237–261.

FAO (2012) Voluntary Guidelines on the Responsible Governance of Tenure of Land, Fisheries and Forests in the Context of National Food Security, FAO, Rome. See: www.fao.org/docrep/016/i2801e/i2801e.pdf.

Fonte, M. (2013) Food Consumption as Social Practice: Solidarity Purchasing Groups in Rome, Italy. *Journal of Rural Studies*, 32, pp.230–239.

Franco, J., Monsalve, S. and Borras, S. (2015) Democratic Land Control and Human Rights. *Current Opinion in Environmental Sustainability*, 15, pp. 66–71.

Ghini, G. (2012) *La Riforma Agraria in Italia e nella Maremma tosco-laziale: cenni storici, Arsial*. Paper Issued in Occasion of the www.agristoria.it website launch, Rome, 5th July 2012.

Gioia, M. and Mari, F. (eds) (2012) Il valore della terra un contributo alla conoscenza del mercato italiano dei terreni agricoli attraverso i dati della RICA, INEA. Available at: https://rica.crea.gov.it/download.php?id=938.

Grando, S., Carey, J., Hegger, E., Jahrl, I. and Ortolani, L. (2017) Short Food Supply Chains in Urban Areas: Who Takes the Lead? Evidence from Three Cities across Europe. *Urban Agriculture and Regional Food Systems*, 2(1), pp. 1–11. doi:10.2134/urbanag2016.05.0002.

Hamilton, W., Bosworth, G. and Ruto, E. (2015) Entrepreneurial Younger Farmers and the "Young Farmer Problem" in England. *Agriculture and Forestry*, 61(4), pp. 61–69.

International Planning Committee for Food Sovereignty (IPC) (2016) People's Manual on the Voluntary Guidelines on the Responsible Governance of Tenure of Land, Fisheries and Forests in the context of National Food Security. A Guide for Promotion, Implementation, Monitoring and Evaluation. Available online at: http://foodsovereignty.org

IPES-Food (2018) Breaking Away from Industrial Food and Farming Systems: Seven Case Studies of Agroecological Transition. http://www.ipes-food.org/_img/upload/files/CS2_web.pdf

Kay, S., Peuch, J. and Franco, J. (2015) Extent of Farmland Grabbing in the EU. Study for the European Parliament's Committee on Agriculture and Rural Development. Available at http://www.europarl.europa.eu/RegData/etudes/STUD/2015/540369/IPOL_STU(2015)540369_EN.pdf.

Leonard, B., Kinsella, A., O'Donoghue, C., Farrell, M. and Mahon, M. (2017) Policy Drivers of Farm Succession and Inheritance. *Land Use Policy*, 61, pp.147–159.

Mela, G., Longhitano, D. and Povellato, A. (2012) *The Evolution of Land Values in Italy. Does the Influence of Agricultural Prices Really Matter?* Paper prepared for the 123rd EAAE Seminar Price Volatility and Farm Income Stabilisation Modelling Outcomes and Assessing Market and Policy Based Responses. Dublin, 23-24 February 2012

Mok, H.-F., Williamson, V. G., Grove, J. R., Burry, K., Barker, S. F. and Hamilton, A. J. (2014) Strawberry Fields Forever? Urban Agriculture in Developed Countries: A Review. *Agronomy for Sustainable Development*, 34(1), pp.21–43.

Moore, M.-L., Riddel, D. and Vocisano, D. (2015) Scaling Out, Scaling Up, Scaling Deep. Strategies of Non-profits in Advancing Systemic Social Innovation. *The Journal of Corporate Citizenship*, 58, pp.67–84.

Salvati, L. (2016) Neither Urban Nor Rural: Urban Growth, Economic Functions and the Use of Land in the Mediterranean Fringe, in Andersson, K., Sjöblom, S., Granberg, L., Ehrström, P. and Marsden, T. (eds) *Metropolitan Ruralities (Research in Rural Sociology and Development, Volume 23)*, Bingley, UK: Emerald Group Publishing Limited, pp.19–37.

Swinburn, B., Egger, G. and Raza, F. (1999) Dissecting Obesogenic Environments: The Development and Application of a Framework for Identifying and Prioritizing Environmental Interventions for Obesity. *Preventive Medicine*, 29(6), pp.563–570. doi:10.1006/pmed.1999.0585.

Van der Ploeg, J. D., Franco, J. and Borras, S. M. (2015) Land Concentration and Land Grabbing in Europe: A Preliminary Analysis. *Canadian Journal of Development Studies*, 36(2), pp.147–162.

Van der Ploeg, J. D., Jingzhong, Y. and Schneider, S. (2012) Rural Development Through the Construction of New, Nested, Markets: Comparative Perspectives from China, Brazil and the European Union. *Journal of Peasant Studies*, 39(1), pp.133–173.

Zagata, L. and Sutherland, L. A. (2015) Deconstructing the 'Young Farmer Problem in Europe': Towards a Research Agenda. *Journal of Rural Studies*, 38, pp.39–51.

Zasada, I. (2011) Multifunctional Peri-Urban Agriculture—A Review of Societal Demands and the Provision of Goods and Services by Farming. *Land Use Policy*, 28(4), pp.639–648.

Žlender, V. and Ward Thompson, C. (2017) Accessibility and Use of Peri-Urban Green Space for Inner-City Dwellers: A Comparative Study. *Landscape and Urban Planning*, 165, pp.193–205.

5

URBAN AGROFORESTRY AS A STRATEGY FOR ALIGNING AGROECOLOGY WITH RESILIENCE PLANNING INITIATIVES

Sarah Lovell and John Taylor

1. Background

Despite growing interest in urban agriculture and visionary proposals for its implementation, we have yet to see broad integration of food production into the urban environment (Lovell, 2010; Lovell and Taylor, 2013). As an alternative to the typical annual vegetable production that dominates the dialogue and the applications of urban agriculture, an 'urban agroforestry' (UAF) system might better align with greening initiatives and help to overcome barriers to adoption. Agroforestry can be defined as the integration of trees and shrubs with crops and/or livestock (e.g., windbreaks protecting sensitive species), but the concept can also encompass the use of trees and shrubs that are themselves crops. Mixtures of different species of nut or fruit trees, berry shrubs and other plants become 'multifunctional woody polycultures' that supply food and a broad array of other ecosystem services that are well documented in diverse perennial plantings (Lovell et al., 2017; Kowalski and Conway, 2019).

Several variations on this concept are emerging in the scientific literature and popular culture. *Permaculture*, a portmanteau of 'permanent agriculture', encourages purposeful design of landscapes to reflect the patterns found in nature and to produce food or other useful materials (Ferguson and Lovell, 2014). *Edible forest gardens* are layered polycultures of perennial woody and herbaceous plants that often facilitate the growth of one another (Jacke and Toensmeier, 2005). Food-producing species are integral to the *food forest* or *urban food forestry* concepts found in the literature, and these systems are sometimes encouraged in order to improve the 'sustainability or resilience of urban communities' (Clark and Nicholas, 2014; Park et al., 2018).

If a UAF approach is used as an advanced form of urban agriculture, the productive species might be better integrated with other urban planning initiatives that promote sustainability, multifunctionality and resilience (Clark and Nicholas, 2014). UAF is potentially more sustainable than conventional urban landscapes such as ornamental gardens, vegetable gardens and lawn, particularly if inputs are managed appropriately (Beck et al., 2001). Linking to nutrient recovery through 'new sanitation' strategies could further reduce the need for outside inputs (Wielemaker et al., 2018). In terms of land access and tenure, the perennial nature of the tree and shrub species would give more permanence and a stronger sense of place that might help to protect the plantings from development or competition from other land uses (O'Brien, 2006). In most contexts, people tend to prefer landscapes that include trees over those that do not (Sullivan et al., 2004; Kuo et al., 1998); this would engender community support. UAF could also be integrated into city parks and other public spaces to better secure the long-term commitment to these plantings. By aligning zoning and tax policies in cities to allow a production component, the potential for edible landscaping could be normalised (McLain et al., 2012).

Soil contamination, another challenge or barrier to urban agriculture, would also be minimised using UAF that introduces trees and shrubs that produce fruits and nuts above the contaminated media. Even in areas with known heavy metal contaminants in the soil, the concentrations found in the edible portions (fruits/nuts) are relatively low and may not be tightly correlated with the levels found in the soil environment (Samsoe-Petersen et al., 2002). The potential to accumulate heavy metals and other contaminants varies by crop species (Gori et al., 2019), so caution must be taken and more research is needed. In areas with heavy contamination, non-edible materials could be produced, such as textiles, dyes and other products (Robinson et al., 2017).

Finally, UAF offers an advantage over annual forms of urban agriculture due to the unique synergy with other urban resilience initiatives. Urban trees, as key components of green space, can help to modulate extreme weather events and disturbances, such as flooding, strong winds and heat waves (US Forest Service, 2016), increasingly associated with climate change. Urban trees can also mitigate climate change by storing carbon (Lehmann et al., 2019) and, as an alternative landscape type, reducing greenhouse gas emissions from conventional land management practices requiring greater inputs of fertiliser and maintenance activity (e.g., lawn mowing) (Lwasa et al., 2015). By focusing on productive trees and shrubs, UAF can be highly compatible with other tree canopy initiatives such as 'ReForest London' (the Million Tree Challenge) (Salbitano et al., 2014). These approaches have been replicated across multiple cities to combat the effects of global warming.

2. Agroecological considerations for urban areas

The integration of UAF into an urban setting offers exciting opportunities to build capacity for the production of healthy fruit and nuts. The design of these UAF systems could be improved by including agroecological principles. Below we review the inclusion of such principles in land planning and site selection, planting design and species selection.

Land planning and site selection

In many urban areas, built development is prioritised over green space, particularly compared with land uses such as urban agriculture, which can be seen as transgressive and as inefficient for valuable urban land. UAF, however, could be intentionally integrated into the existing or planned urban landscape. A number of studies demonstrate that food foraging across the urban environment is already a widespread practice, and that productive edible species often grow, though hidden to many residents, amidst parks, forests, cemeteries, schoolyards and even street tree plantings (Landor-Yamagata et al., 2018; Shackleton et al., 2017). Residents partake in this activity for many important reasons, including valuing the food, gaining enjoyment, satisfying curiosity, improving their health and connecting with nature (Landor-Yamagata et al., 2018; Synk et al., 2017). They forage in cities across the globe, from high- to low-income countries (Hurley and Emery, 2018; Landor-Yamagata et al., 2018; Shackleton et al., 2017), demonstrating that, in some sense, UAF is already an integrated system, although often not an intentionally planned one. To further expand this practice, residents' awareness of the edible species that surround them needs to be raised through education, and policies need to be adjusted to allow harvesting of fruit, nuts, or understorey edible plant products in areas where it is safe to do so (Shackleton et al., 2017).

However, to reach even more residents and to make a greater contribution to urban production, additional land must be converted to new UAF. Such land may be found in 'marginal' and interstitial spaces with limited potential for commercial development (Bjorklund et al., 2019). Marginal lands can include:

(1) parcels that are vacant due to neighbourhood disinvestment;
(2) sites that are unbuildable due to topographic features, size, or shape;
(3) zones such as rights-of-way that are restricted from building because of utilities or infrastructure;
(4) areas that are prone to flooding (Betz et al., 2016).

Land inventories for urban agriculture, however, have mostly been developed based on criteria for growing annual vegetables, as opposed to perennial food forests. In Madison, WI, USA, for example, the inventory was defined based on criteria such as minimum size lot, limited slope, water access, solar access and avoiding conflicting land uses such as recreational areas (Eanes and Ventura, 2015). Solar irradiance was the primary criterion for identifying potential food-producing land in the Seattle foodshed. These criteria, however, might be unnecessarily limiting for UAF, which includes varied forms. Sloped land, for example, would be suitable for mixed plantings of trees, shrubs and understorey species that could supply food for foragers while preventing soil erosion and creating a wildlife habitat. Shade need not preclude food production; forest gardens of shade-tolerant edible and medicinal species could be developed beneath the existing urban tree canopy. If fruit production is a key desideratum, then sites with adequate sunlight (preferably at least 8 hours/day) should be prioritised. While a number of woody plant species are adapted to low light conditions, their fruit production declines substantially when sunlight is limited, particularly in temperate regions.

As noted previously, soil contamination can be a limiting factor, particularly in marginal areas. Exposure to contaminants at the time of planting can be an additional concern, although soil disturbance is much reduced with the perennial woody and herbaceous plants that make up UAF, compared to the annual (or more frequent) disturbance required to establish conventional vegetable crops. The potential for exposure of the site to airborne contaminants should also be a consideration (Gori et al., 2019).

Potential synergies with adjacent, complementary, or similar land uses should also be considered when siting UAF. Colding (2007) proposes the concept of 'ecological land-use complementation' to encourage connecting green spaces to increase the benefits for biodiversity conservation and ecosystem services. Aggregating and connecting UAF sites across property boundaries could have ecological benefits. It could also enhance system productivity by increasing total habitat area, reducing edge effects and facilitating the movement of species, including resource species providing services to production, across the otherwise fragmented urban landscape. Managers of adjacent sites could engage in shared learning (Cabell and Oelofse, 2012) in the management of their UAF systems, and share material resources such as irrigation water for the establishment of new plantings (Robinson et al., 2017). UAF could also potentially benefit from being strategically located near built sites. Rainwater harvested from rooftops, for example, could be used to irrigate plantings. By blocking winter winds, buildings may create microclimates conducive to the growth of marginally hardy species. Building residents can serve as UAF stakeholders helping to manage and ensure the conservation of sites while reaping their productive benefits.

Planting design and arrangement

Guidelines for planting design and arrangement can be based on the multi-strata systems that are promoted in permaculture and other food forest literature. A survey of community orchards in the United States found the most common management style to be described as 'permaculture' (Betz et al., 2016). These approaches all share the goal of creating an intentional landscape based on the complex structure of a forest but using component plant species that yield food for human consumption, as demonstrated by a food forest in Kansas City, MO, USA (Figure 5.1). The various vertical layers can include:

FIGURE 5.1 Food forest on less than 0.1 ha of land, maintained by Cultivate KC in Kansas City, MO. Fruits on right include, from top to bottom, Kousa dogwood (*Cornus kousa*), jujube (*Ziziphus jujube*), American persimmon (*Diospyros virginiana*) and pawpaw (*Asimina triloba*).

(1) tall tree canopy with nut and fruit trees;
(2) low tree layer with small or dwarf fruit trees;
(3) shrub layer with berry-producing species;
(4) herbaceous plants, including herbs and vegetables;
(5) ground-cover plants for soil stabilisation;
(6) root crops with harvestable component below the soil;
(7) vining/climbing species that depend on trees or other structures for vertical growth (Jacke and Toensmeier, 2005).

Edible mushrooms can also be grown on the surfaces of lower layers in the shaded canopy. Honey-producing bees can be included as another productive component with the added benefit of serving as pollinators.

The benefits of this complex planting structure are many. Lehmann et al. (2019) found that both the trees in the overstorey and the woody plants in the understorey contribute to carbon sequestration (Lehmann et al., 2019). The high diversity of plants likely contributes to biodiversity at higher trophic levels, including arthropods that can provide pollination and natural pest suppression services (Bennett and Lovell, 2014, 2019). Plant diversity can also diversify the diets of urban residents (Ickowitz et al., 2014).

While UAF sites would ideally include all of these layers, system simplification may be necessary depending on management goals. These goals could include reducing maintenance needs (a particular concern in low-income neighbourhoods) or prioritising the production of certain highly valued foods. The top canopy layer may be omitted to allow greater penetration of sunlight into the lower layers, a strategy that can also optimise functions for a planting that is space-limited. In Kinshasa in the Democratic Republic of the Congo, multipurpose trees are often planted at the borders of gardens, in some cases purposefully providing shade for other crop species (Etshekape et al., 2018). Using woody plants in the border areas of urban agriculture sites could provide multiple benefits, even for the annual plants growing in the interior. The woody plants could provide a buffer from hot, dry winds, creating a microclimate conducive to the growth of understorey vegetation, and assist in infiltration of stormwater that might otherwise flood or run off the site (Wortman and Lovell, 2013).

Regardless of the complexity in vertical layers or the location of plantings, an appropriate design will use a successional approach, recognising that the system will change over time as trees and shrubs grow and mature. Available light in the understorey will decrease, so the selection of plants for that zone must evolve to match the new conditions. Such a successional approach may also shape the maintenance and management of the site. Intensive management during system establishment (Park et al., 2018), however, pays off over the long term. Landscape architects have little consideration for managing succession and often fail to acknowledge time as a factor in their planting designs. Moreover, most landscape management practices, e.g., mowing, seek to suppress succession. For UAF to become a common component of cities, a paradigm shift will be needed in urban design theory, education and practice.

Species selection

The use of a multi-strata, diverse planting design allows for the inclusion of a wide range of edible and supporting species (e.g., nitrogen-fixing plants). Several key criteria for selection apply. Species should: (1) be well-adapted to the local climate and urban environment, (2) be low maintenance and low input, (3) have a good capacity to promote human health and (4) be culturally acceptable and, moreover, culturally preferred.

In order to be 'well-adapted', the temperature, precipitation, and sunlight requirements for plant species need to match those of the local climate. Given the focus on perennial plants, this needs to be within a long-term perspective of future climate change. These conditions can be further influenced by the urban context, including diverse microclimates but generally characterised by elevated temperatures (i.e., the urban heat island effect) and increased vapour pressure deficits. Sunlight may also be limited when built structures cast shadows on a site. Also, while irrigation may be available to supplement precipitation, the use of treated drinking water as an irrigation source is not ideal; species should be either adapted to local precipitation regimes or sustainable sources of water, such as rooftop runoff, should be used.

Even given these restrictions, a wide range of edible species meet selection requirements and can potentially thrive in many urban areas, as demonstrated by the large number of species found in studies of urban foraging. A study of New York City found 119 species of street trees and 201 total species that could provide food and contribute to a culture of 'urban foraging' (Hurley and Emery, 2018). In Berlin, a study found that 125 plant taxa were foraged, 16% from trees and 13% from shrubs.

Clark and Nicholas (2014) developed a Climate-Food-Species matrix to consider suitability for the city of Burlington, VT, USA, a city with a humid continental climate with cold winters. Examples of highly suitable species included *Vaccinium* sp., *Malus domestica*, *Prunus* sp. *Pyrus* sp., *Rubus* sp., *Ribes* sp. and others (Clark and Nicholas, 2014). While the list was broad, additional species might rate higher if the criteria item of 'widely recognized and marketed' were to be ignored, allowing new markets to develop around this specific form of food production.

The choice of well-adapted plant species will likely naturally result in the selection of plants with low maintenance and low input requirements. In urban environments, the use of pesticides as an agricultural input is of particular concern because of the heightened potential for human contact or ingestion. Concerns related to pesticide use have, in fact, been used to justify regulations prohibiting foraging or other forms of urban food production, in some cases with good reason (Shackleton et al., 2017). Species and cultivars should be selected for their disease and insect resistance, to reduce the need for pesticides.

Plant selection should emphasise human health dimensions. Poe et al. (2013) suggest moving beyond 'food justice' to consider 'health justice' with foods for good nutrition and medicinal plants for remedies and ailments (Poe et al., 2013). Some of the most nutrient dense foods that are high in antioxidants grow on shrubs and trees adapted to a variety of conditions and climates (e.g., *Aronia* sp., *Sambucus* sp.). Plants should also be selected for their ability to improve the quality of urban life by alleviating discomfort from heat, filtering air pollutants, reducing noise, encouraging physical activity and improving psychological well-being (Salbitano et al., 2014). For all plants, hazard minimisation should be considered from a human health perspective. For example, use of large-fruited species in areas of high pedestrian traffic could potentially result in injury to walkers (Shackleton et al., 2017).

Multiplying the benefits

UAF can offer a wide variety of benefits for the health of the environment and its human inhabitants. The case for full integration into the urban context could be made stronger, however, if these specialised green spaces could generate a profit (Schaffer et al., 2019). The financial viability of urban agriculture projects, when all labour and other inputs have been accounted for, is contentious (Dimitri et al., 2016). Research on the economic sustainability of UAF is scant. In a study on the profitability of existing examples of UAF in Villa El Salvador, Peru, Lafontaine-Messier et al. (2016) found that initial investments could be recovered in four to eight years, but only if the cost of labour was not considered.

To improve financial performance, the authors proposed valuing UAF's educational and other functions, relying on labour already paid for through city funds for urban tree maintenance, or bringing in volunteers (Lafontaine-Messier et al., 2016). Tree nurseries or landscaping businesses focusing on edible landscaping could be profitable models (Carsan et al., 2010). Robinson et al. (2017) estimated that edible landscaping businesses in Phoenix, AZ could yield US$4 to US$66 billion in additional value for the metro area, depending on the percentage of land converted to UAF (range: 10–35%). Edible landscaping could further contribute to the local economy by creating new jobs through specialised landscaping, nursery products and urban farmers (Robinson et al., 2017).

Apart from their economic integration, there is great opportunity to align UAE with existing strategies in public policy, in particular the integration of Tree Canopy Plans. Planting more trees is considered to be one of the most effective strategies for adapting to and mitigating climate change (Bastin et al., 2019; Nowak and Crane, 2002), and many cities have developed specific targets for increasing their tree canopy cover and species diversity (Ordonez and Duinker, 2013). The 'Million Trees' programme, started in the United States, has been adopted by New York City, Los Angeles, Miami and others to recognise the value of urban forests' wide range of ecosystem services.

UAF could be aligned with these and other ecologically based planning goals such as stormwater mitigation and microclimate control (Clark and Nicholas, 2014; Lwasa et al., 2015). Pairing UAF with these programmes might be a simple matter of education and training, offering design guidelines and plant species recommendations. Unfortunately, these opportunities are rarely captured. In a focused study of 47 municipalities across Canada, fewer than a third included fruit trees (edible for humans) in policy initiatives or guidelines. Municipal regulations may reflect a lack of appreciation of UAF as a multifunctional land use (Miljković and Crnčević, 2019). In some cities, such as Seattle, guidelines from civic departments such as Parks & Recreation and the Department of Transportation discourage growing or eating edible fruit or nuts from UAF (McLain et al. 2012).

Additional research would help to make the case for integrating food-producing trees. Much of the information needed to make informed decisions on this subject is already available in the fields of agroforestry, plant science and horticulture. In the following case study of Chicago, IL, the potential for the broad adoption of UAF is considered.

3. Exploring opportunities for the integration of UAE – the Chicago case

Diverse actors promote food production in Chicago and other US cities for its multifunctionality; however, the urban agriculture projects they promote often reproduce at smaller scales the form and, to some extent, the philosophy and practices of conventional farming. Their reliance on annual crops and exogenous sources of nutrients and other inputs, even if organic, undermine their sustainability and resilience. In these schemes, the presence of perennial food plants is often reduced to the occasional orchard or berry planting added as an afterthought to an urban farm or community garden. We argue for a more radical vision of urban agroforestry for Chicago that goes beyond these timid efforts to introduce woody food plants into urban spaces, one that weaves multistorey food forests into the fabric of the city, transforming it ecologically, socially, and politically. Historically, the City of Chicago has promoted parks and other open spaces for the health of the public and the environment as well as for their beauty. The planning and preservation of these spaces has created a spatial framework that can serve as the backbone for a multivalent, transformative urban agroforestry system extending from the scale of the large forest preserves in the Cook County suburbs to the scale of the backyard garden in the urban core.

Chicago is the third most populated city in the United States, with 2.7 million residents within the city limits, 2.5 million more in the adjoining suburbs of Cook County and 9.5 million in the entire Chicago-Naperville-Elgin metropolitan statistical area spanning three states. The city's population is 29.0% Hispanic/Latino, 32.8% white alone (non-Hispanic), 30.1% African American alone and 6.4% Asian alone; 2.7% of Chicagoans identify as two or more races (https://www.census.gov). The city experiences a humid continental climate with four distinct seasons, despite some moderation of temperatures provided by Lake Michigan. Average temperatures are 15°C high and 5°C low, and average annual precipitation is 95.9 cm, with annual snowfall of 84.8 cm (NOAA, 2013).

The City of Chicago's commitment to the conservation and development of urban green infrastructure dates to its very beginnings. At the time of its incorporation in 1837, city leaders adopted the motto *Urbs in Horto,* or City in a Garden (City of Chicago, 2009). By the end of the 19th century, an extensive network of parks and open spaces connected by a boulevard system had developed across the city. Burnham and Bennet's 1909 *Plan of Chicago* (the 'Burnham Plan') further encouraged a green infrastructure approach, catalysing the preservation of the lakefront for recreational purposes, extending the boulevard system and driving the expansion of the Cook County Forest Preserves. Today, the Chicago Park District boasts over 8,100 acres of open space in 580 parks, 50 natural areas and 2 conservatories (http://www.chicagoparkdistrict.com/). These spaces are a key layer of the city's green infrastructure as shown in Figure 5.2. On a regional scale, the Forest Preserve District of Cook County manages more than 68,000 acres of land found mostly in the suburban areas just beyond the boundaries of Chicago. These larger holdings make an important contribution to the conservation of land and wildlife, while also connecting people with nature (fpdcc.com).

Current green infrastructure initiatives

Few green infrastructure initiatives promulgated by regional government or non-government organisations explicitly recognise forested agroecosystems, or

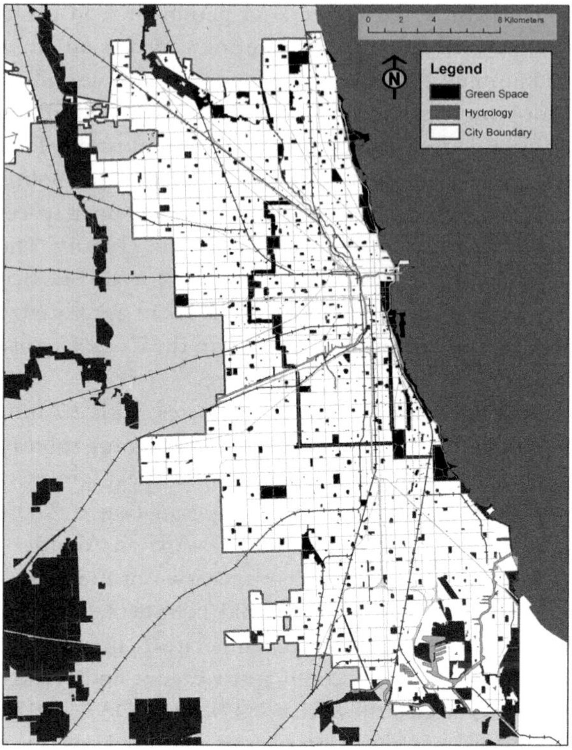

FIGURE 5.2 Green spaces in Chicago, including natural habitats, forest preserves, city parks, boulevards, and campus parks – all designated as solid black. Map created in ArcGIS by Sarah Lovell.

any type of agroecosystem, as a form of green infrastructure. However, the existing initiatives demonstrate a strong interest in networked green infrastructure and could, with adaptation, serve as a foundation for building an interconnected system of agroforestry projects in the city and beyond.

Chicago Wilderness: This is an alliance of over 250 organisations committed to the protection and restoration of regional natural areas; it has developed a Green Infrastructure Vision (GIV). Their vision represents a biodiversity plan for adding and connecting an additional 1.8 million acres of conserved habitat to the current 370,000 acres of protected natural habitat areas in a four-state region encompassing northeastern Illinois, southwestern Wisconsin, northwestern Indiana and southwestern Michigan. While neither the vision nor the biodiversity plan explicitly recognises either agriculture or agroforestry as green infrastructure, the narrative accompanying the map proposes areas for conservation development in which habitat conservation, urban expansion and open space development go hand in hand.

Openlands (openlands.org): This is a conservation organisation founded in 1963. It similarly advocates for conserving and connecting natural and open spaces in the Chicago metropolitan region, and has contributed to the conservation of 55,000 acres of green space since its founding. It includes the creation of school gardens (57 so far) and the greening of school yards. Unlike Chicago Wilderness and other regional initiatives, Openlands includes agroecosystems as part of its planning agenda.

Chicago's Urban Forest Agenda (City of Chicago, 2009): This advocates for multifunctional green infrastructure by quantifying and expanding tree cover throughout the city. The agenda recognises the many ecosystem services provided by urban forests and other green infrastructure. In particular, it highlights the important role of tree cover in managing urban heat islands, thereby moderating energy costs and reducing heat-related deaths of urban residents.

The Chicago Climate Action Plan (http://www.chicagoclimateaction.org/) is guided by the goal 'to reduce our emission and prepare for change' through a wide range of strategies including implementing landscape features for environmental cooling and rainfall capture, preserving plants and trees with a focus on species that thrive in an altered climate, and educating the public about green landscaping.

The Chicago Metropolitan Agency for Planning (CMAP) (http://www.cmap.illinois.gov/2040/): This is the official regional planning organisation for the seven counties of northeastern Illinois. The organisation has developed the GO TO 2040 and GO TO 2050 comprehensive plans to promote sustainable development in the Chicago region. The former explicitly seeks to promote urban agriculture on vacant or underutilised land, to preserve existing farmland for food production and to increase food access to eliminate food apartheid. The GO TO 2050 plan unfortunately does not address food explicitly but does call for the integration of green infrastructure at multiple scales within a rather narrow definition, focusing on the conservation of 'high-quality' natural areas and stormwater infiltration with production excluded from the list of co-benefits.

The City of Chicago's Green Stormwater Infrastructure Strategy: Published in 2014, this includes measures such as the use of green roofs and permeable streets and alleys to reduce combined sewer overflows during even moderate precipitation events. The strategy's list of co-benefits of a green stormwater infrastructure is long but, again, fails to recognise food production as a potential co-benefit through the capture and use of stormwater for irrigation in urban food gardens and farms, or the interception of precipitation by a productive tree canopy.

While regional and city-scale organisations can support agroforestry projects by providing vital informational, material and financial resources, projects must reflect the values and needs of local communities. Ensuring representation in planning and equity in the distribution of benefits of agroforestry projects will require engaging partners at the neighbourhood level. Three neighbourhood level initiatives in particular hold the potential to partner on urban agroforestry.

Grow Greater Englewood is currently collaborating with NeighborSpace, a Chicago community land trust, to develop community-controlled farms along a repurposed elevated rail line as part of the Englewood Land Access Project in the Englewood Urban Agriculture Zone. The project is designed to address a lack of access to food and jobs, and high rates of poverty and food insecurity in the Englewood community area on the city's South Side. Once a thriving neighbourhood, one-fifth, approximately 700 acres, of the neighbourhood is now vacant land that could be redeveloped for gardens, parks, cultural activities and affordable housing. Management models for the planned farms include for-profit operations and cooperative farming enterprises.

Little Village Environmental Justice Organization (LVEJO) has worked for over 25 years for environmental justice in Little Village, a majority Mexican-American neighbourhood in South Lawndale on the city's South Side. The organisation was originally founded in response to the potential exposure of children to paint containing lead particles during the renovation of a local public school. The organisation's current mission is 'to build a sustainable community that promotes the healthy development of youth and families, provides economic justice and practices participatory democracy and self-determination' (LVEJO, 2019). As part of its mission, LVEJO founded a 0.6-ha community garden, the Semillas de Justicia garden and spearheaded the redevelopment of a 54.3 EPA Superfund site as a public park, La Villita Park. Although the community garden focuses on annual crops and the park is planted primarily with the typical trees in lawn pastoral style typical of many of Chicago's parks, the community could look to traditional forms of agroforestry indigenous to Mexico, such as the *milpa*, for inspiration for urban agroforestry in public and private spaces. Local adaptations of these traditions may already be present in Chicago. A study of home food gardening in the city found that fruit trees were much more abundant and made a significantly larger contribution to the production area of the gardens of Mexican-origin households compared to other cultural groups (Taylor et al., 2017).

Blacks in Green was founded to catalyse the 'the design and development of green, self-sustaining, mixed-income, walkable-villages in communities owned and populated by African American' (http://blacksingreen.org/). A concept key to the walkable-village is the 'homestead', the combination of affordable green homes and gardens producing healthy food. While the nature of these gardens is nascent, urban agroforestry could provide sustainable models for the homestead and for community-based food production, including agroforestry models rooted in the rich traditions of African polycultural agroecosystems.

Future opportunities for urban agroforestry

We envision a networked, multifunctional forested agroecosystem encompassing Chicago's public and private lands, from parks to residential lots. Different strategies will be needed to implement this vision at each scale.

Introduction of agroforestry in Chicago's urban parks. To begin with the low-hanging fruit, while Chicago's extensive park district encompasses 576 separate parks, the extent of food-producing urban agriculture is limited to 30 sites. The extent of agroforestry, despite the forested nature of many parks, is unknown but likely to be negligible. To consider the potential for expanding functions in parks, we conducted an analysis of the Chicago park system to characterise the individual parks based on size. A Geographic Information System (GIS) was assembled based on spatial data layers acquired, analysed and processed in ESRI® ArcMap™ version 10. Landscape analysis of Chicago parks was based on a Shapefile of all parks managed by the Chicago Park District, and K-Cluster analysis was performed in SPSS based on park size [transformed as: log10(area)] to classify the existing parks (see Table 5.1).

While very large parks were considerably fewer in number ($N = 30$), their size (64 ha average) makes them a huge contributor to the overall park area, providing 1,910 ha or 67% of the total 2,842 ha of park land.

Agroforestry systems could be strategically introduced into parks. The exact nature of each intervention would depend on the park's size and ecological character and the interests and needs of the surrounding community (Table 5.2). In some parks, multifunctional food forests, multi-layered systems of food-producing herbaceous perennials, shrubs, vines and small and large trees (Park et al., 2019) could be created to provide food, recreational opportunities and habitat for wildlife, to enhance park aesthetics, to moderate the urban heat island effect and to infiltrate stormwater. In other contexts, simplified agroforestry systems, such as orchards with herbaceous annual or perennial ground layer crops, might be more appropriate due to the interests and needs of local residents or to address safety concerns about the reduced sightlines in multi-layered systems.

Repurposing vacant lots and underused sites and redesigning residential gardens. Vacant lots may offer one of the most obvious opportunities to expand urban agroforestry, particularly in under-resourced neighbourhoods. The city owns many vacant lots through tax lien acquisition following the city-paid demolition of the

TABLE 5.1 Results of cluster analysis of Chicago's 576 city parks based on size

Size/Class	*Number* (n)	*Mean Area (ha)*	*Min Area (ha)*	*Max Area (ha)*
Small	180	0.112	0.008	0.237
Medium	191	0.690	0.244	1.493
Large	174	4.485	1.533	11.502
Very Large	30	63.666	13.442	401.198

Table created by authors

TABLE 5.2 Potential production functions that could be integrated into various forms of urban green space based on size and habitat

Habitat/Feature	*Min park size*	*Considerations*	*Production functions*
Native woodland	Large	Low disturbance	Collection of fallen nuts or gathering of understorey berries, mushrooms, etc.
Restored prairie	Medium	Lacks woody plants, Seasonal burning needed	Supports pollinators that can increase production capacity of nearby species
Wetland	Large	Water retention	Native species that produce stems for floral cuttings
Raingarden	Small	Stormwater capture, possible contaminants	Adapted edible plant species can be included if contamination potential is low
Community garden	Large	Visual quality; theft	A greater diversity of edible species can be encouraged, with orchards and berry hedges
Edible landscaping	Small	Fruit drop; maintenance	Edible species selected for low maintenance and good ornamental character
Native plantings	Small	Urban adaptability	Native species that produce edible fruits and nuts encouraged

buildings on site. In neighbourhoods with demand for residential or commercial development, the lots can be reused for built infill, but in other neighbourhoods, they might best be repurposed as multifunctional green infrastructure. Through the City of Chicago's Adjacent Neighbors Land Acquisition Program (ANLAP), homeowners in certain parts of the city can purchase a vacant lot adjacent to their property for as little as 10% of the lot's fair market value (City of Chicago, 2019a). Participants in ANLAP must improve the lot within six months of purchase; specified forms of improvement include landscaping, which could be in the form of forest gardens.

Contiguous vacant lots, however, might be repurposed for a community forest garden or for forest farming with a full complement of tree layers. Historically, city-owned vacant lots have been conveyed to community land trusts for community gardens managed by local residents. Urban agriculture organisations have also used city-owned vacant lots in usufruct. The need for long-term land tenure for agroforestry projects, though, would preclude the latter option. There are specific legal barriers that may limit the size of residential lot-based agroforestry initiatives. Outside Public Open Space zones, where no areal limitations apply, community gardens are limited to 2,323 m^2 (25,000 ft^2) in size, the equivalent of eight standard city lots, to prevent a single garden from dominating a block. A single block, though, may have more than one garden of the maximum size. While the maximum size of farms is not capped, they are prohibited in all residential and some business zones and require a business licence and, as a business, off-street parking (City of Chicago, 2019b).

For mapping city-owned vacant lots, a list of properties with addresses was obtained from the city's website (http://www.cityofchicago.org/). The addresses were geocoded to create a separate layer for mapping in ArcMAP, together with the community area boundaries (Figure 5.3). At the time of the analysis, the city listed 11,260 vacant parcels for a total area of 575 ha, with parcel sizes averaging 510 m^2 and median of 291 m^2. (This inventory does not include the many more vacant lots not owned by the city.) Clearly, the parcels are unevenly distributed and largely concentrated in community areas on the West and South Sides with

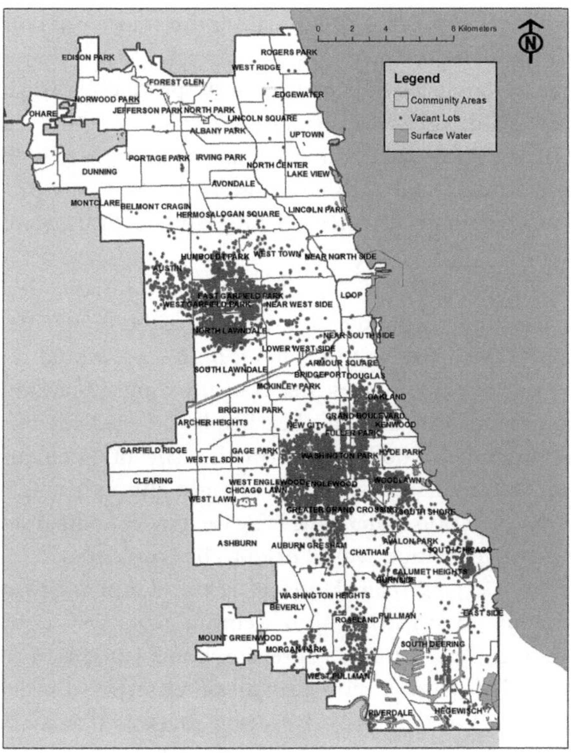

FIGURE 5.3 Map of Chicago's city-owned vacant lots, with a dot to indicate each individual parcel.

majority African American populations, reflecting the city's continuing history of segregation and disinvestment. Establishing community-driven urban agroforestry projects on even a fraction of these lots could begin to help to address deeply entrenched issues of poverty, income inequality, food apartheid and disparities in environmental quality.

The remediation and revitalisation of brownfields and quarries offers another opportunity to expand urban agroforestry in the Chicago region. Cook County has the highest proportion of brownfield sites in the state. These have been the focus of repeated redevelopment plans, including the Chicago Brownfields Initiative (City of Chicago, 2003) and, more recently, a collaboration between the Delta Institute and LVEJO to develop, with community guidance, strategies for redeveloping two large brownfield sites in the Pilsen and Little Village neighbourhoods (Delta Institute – LVEJO 2016). While the latter plan does not include any outdoor food production, it does include several food-centric strategies: a community kitchen, a food waste composting facility and indoor farms. Given the low-level contamination of brownfield sites, redevelopment strategies for these and other sites could be expanded to include agroforestry options. In the early stages of remediation, woody biofuel crops such as poplar (*Populus* sp.) and willow (*Salix* sp.) might be grown for use as an energy source (Atwell et al., 2010). After remediation, fruit and nut trees with low potential for the transfer of contaminants to edible portions could be grown.

Chicago boulevards could be re-envisioned as continuously productive urban landscapes; this is if they were re-designed to include more than just turf and standard street trees. The existing boulevard system contains approximately 90 ha of unpaved land that might be designed with food forests that not only provide edible products but also contribute to microclimate control, stormwater management and visual quality.

Right-of-way (ROW) zones along highways and rail corridors and under electrical transmission lines are also underutilised spaces that could be repurposed for urban agroforestry. Some ROWs, such as those along highways, would not be appropriate for public use due to safety issues, inaccessibility and unpleasant noises; however, like contaminated brownfields, they might be used for biomass production with short rotational woody crops. In the case of easements under electrical lines, connected open spaces often run right through neighbourhoods and already require vegetation management by the power companies. Limitations on the mature size of vegetation would need to be considered.

Redesigning residential lots for food production. Housing is the dominant form of land use in Chicago, and home food gardening is a common activity across cultural groups (Taylor and Lovell, 2014). In the aggregate, home gardens make a much larger contribution to urban food systems than all other forms of urban agriculture combined (Taylor and Lovell, 2012). Research suggests that many households' food gardens contain little vertical structure in the form of fruiting shrubs or fruit or nut trees (Taylor et al., 2017). In fact, gardeners accustomed to annual cropping systems may view trees negatively because of the negative impact of the shade they cast on their gardens. Residential-scale agroforestry models that address gardeners' concerns and support the production of culturally appropriate foods while providing the co-benefits of forested agroecosystems need to be developed and disseminated. A large percentage of Chicago residents, 20.7% in 2018, are foreign born (US Census Bureau, 2018b).

Expanding research and education on campus greens. In a dense urban area such as Chicago, campus greens at all levels of education are often relatively small and have complex shapes as they fit between and around buildings and parking lots. These sites may not be appropriate for full open access for the public; however, campus greens could offer great opportunities for students to learn about agroecology and urban agroforestry through demonstration gardens that provide experiential learning opportunities as part of the curriculum. The City of Chicago GIS identifies 112 sites as Campus Parks, with an area totalling 109 ha and average size of 0.98 ha. The largest campus park is a 15 ha site at Chicago High School for Agricultural Sciences (CHSAS), where students are involved in summer programming related to food production and entrepreneurship.

Developing a community of urban agroforesters in Chicago

The vast majority of urban farmers and home and community gardeners are unlikely to be familiar with the benefits, principles and practices of urban agroforestry. Agroecological systems are knowledge-intensive, and designing and managing forested agroecosystems can be challenging even for experts. Metropolitan Chicago has multiple educational resources that could be leveraged to train would-be agroforesters and to provide ongoing outreach. The Chicago Botanic Garden's Windy City Harvest programme operates three youth farms within the city and one in suburban Waukegan, four sustainable agriculture training farms and four contract partnership farms in collaboration with other organisations. The programme offers a variety of educational opportunities for amateur and professional urban agriculturalists (Chicago Botanic Garden, 2019) that could be extended to hands-on instruction in urban agroforestry at the programme's farms, although annual crops currently dominate those farms.

The University of Illinois Extension offers outreach to gardeners and farmers on a broad spectrum of agricultural and horticultural topics, including tree and small fruit production (University of Illinois Extension, 2019a). Recognising that peer-to-peer training is the most effective form of outreach, Extension's Master Gardener programme trains interested members of the public to become horticultural educators. Illinois currently has over 3,200 Master Gardeners (University of Illinois Extension, 2019b). To the best of our knowledge, agroecology is seldom covered in Master Gardener training nationwide. One of us (Taylor) has introduced a module on agroecosystems and agroecological practices to the Master Gardener Core Training at the University of Rhode Island, and it has been very well received by trainees. Master Gardeners with an interest in the topic could receive special training to practice and educate the public about urban agroforestry systems, or an independent programme managed by Extension or another organisation could be developed. The programme could be modelled on the Australian Master Tree Grower programme, in which farmers participate in a short course on agroforestry combined with instruction on information and peer network building that enables them to create their own self-supporting networks on agroforestry practitioners (Reid, 2019).

The Center for Agroforestry at University of Missouri in Columbia, MO, offers a variety of methods to learn about agroforestry and the applications for the urban environment. The annual Agroforestry Academy (http://www.centerforagroforestry.org/academy/) is a week-long intensive training programme for farmers, landowners and practitioners. The combination of classroom activities, farm visits and design activities provides participants with the valuable tools needed to develop UAF appropriate for the city of interest. For specific topic areas, the Center for Agroforestry offers a wide range of resources, including webinars, podcasts and outreach publications. A deeper understanding of scientific basis for agroforestry can be found through the online courses offered by the Center for Agroforestry. Individuals who have previously obtained a Bachelor's degree can apply to the online Agroforestry Master's degree or the online graduate certificate.

4. Conclusions

This chapter provides a model for applying multifunctional green infrastructure to a city, in this case, Chicago, IL. The city was built on an area that was once a heterogeneous landscape of prairie, wet prairie, marsh and savannah that supported a high degree of biodiversity. While most of the original ecosystems have been destroyed, the city has evolved with a strong emphasis on urban greening, including a historical focus on networked green space through the park system, boulevards and forest preserves. Several current initiatives seek to expand the green infrastructure of the city through programmes such as Chicago Wilderness Alliance, Urban Forest Agenda, Chicago Climate Action Plan and others. Less attention has been given to the potential for production functions from the existing spaces, so we explored opportunities to add urban agroforestry in public parks, community gardens, vacant and underused sites and campus greens.

Through this review, we demonstrate the potential to improve system performance by building on the historic context of the site, leveraging existing programmes and initiatives, expanding functionality into underused spaces and empowering communities to alter the landscape change trajectory to better meet social and ecological needs. Urban planning strategies that embrace and facilitate such an approach can help cities navigate the inevitable yet unpredictable future challenges related to climate variability and competition for limited resources. We call on planners, policy-makers and researchers to offer thoughtful solutions to support multifunctional green infrastructure and to develop and pursue innovative strategies for actively engaging local communities in the development of that infrastructure, from project initiation to ongoing, community-based management of the completed project. Only a democratic and participatory process can help to ensure that multifunctional green infrastructure development benefits all stakeholders, including those from historically disadvantaged and underrepresented populations.

References

Atwell, R. C., Schulte, L. A. and Westphal, L. M. (2010) 'How to build multifunctional agricultural landscapes in the US Corn Belt: Add perennials and partnerships', *Land Use Policy*, vol 27, no 4, pp1082–1090.

Bastin, J. F., Finegold, Y., Garcia, C., Mollicone, D., Rezende, M., Routh, D., Zohner, C. M. and Crowther, T. W. (2019) 'The global tree restoration potential', *Science*, vol 365, no 6448, pp76–79.

Beck, T. B., Quigley, M. F. and Martin, J. R. (2001) 'Emergy evaluation of food production in urban residential landscapes', *Urban Ecosystems*, vol 5, no 3, pp187–207.

Bennett, A. B. and Lovell, S. T. (2014) 'Comparison of arthropod abundance and arthropod mediated predation services in urban green spaces', *Insect Conservation and Diversity*, vol 7, no 5, pp405–412.

Bennett, A. B. and Lovell, S. T. (2019) 'Landscape and local site variables differentially influence pollinators and pollination services in urban agricultural sites', *Plos One*, vol 14, no 2, p.30212034.

Betz, M., Mills, J. and Farmer, J. (2016) 'A preliminary overview of community orcharding in the United States', *Journal of Agriculture, Food Systems, and Community Development*, vol 7, no 2, pp13–28.

Björklund, J., Eksvärd, K. and Schaffer, C. (2019) 'Exploring the potential of edible forest gardens: Experiences from a participatory action research project in Sweden', *Agroforestry Systems*, vol 93, no 3, pp1107–1118.

Cabell, J. F. and Oelofse, M. (2012) 'An indicator framework for assessing agroecosystem resilience', *Ecology and Society*, vol 17, no 1, p13.

Carsan, S., Osino, D., Opanga, P. and Simons, A. J. (2010) *Urban Agroforestry Products in Kisumu, Kenya: A Rapid Market Assessment*. New York: Springer.

Chicago Botanic Garden (2019) https://www.chicagobotanic.org/adult_education/windy_city_harvest. Accessed 18 December 2019.

City of Chicago (2003) 'Chicago brownfields initiative - Recycling our past, investing in our future', https://www.csu.edu/cerc/documents/ChicagoBrownfieldsInitiativeRecyclingOurPastInvestinginOurFuture.pdf. Accessed 1 May 2020.

City of Chicago (2009) *Chicago's Urban Forest Agenda*. Chicago, IL.

City of Chicago (2019a) 'Planning and development - Adjacent neighbors land acquisition program (ANLAP)', https://www.chicago.gov/city/en/depts/dcd/supp_info/adjacent_neighborslandacquisitionprogramanlap.html. Accessed December 18, 2019.

City of Chicago (2019b) 'Planning and development - Urban agriculture FAQ', https://www.chicago.gov/city/en/depts/dcd/supp_info/urban_agriculturefaq.html. Accessed 18 December 2019.

Clark, K. H. and Nicholas, K. A. (2014) 'Introducing urban food forestry: A multifunctional approach to increase food security and provide ecosystem services', *Landscape Ecology*, vol 28, no 9, pp1649–1669.

Colding, J. (2007) 'Ecological land-use complementation for building resilience in urban ecosystems', *Landscape and Urban Planning*, vol 81, nos 1–2, pp46–55.

Delta Institute - LVEJO (2016) 'Little village vacant property and brownfield redevelopment strategy', https://delta-institute.org/2016/03/nonprofit-collaboration-aims-to-transform-little-village-brownfields-into-sustainable-business-opportunities/. Accessed May 1, 2020.

Dimitri, C., Oberholtzer, L. and Pressman, A. (2016) 'Urban agriculture: Connecting producers with consumers', *British Food Journal*, vol 118, no 3, pp603–617.

Eanes, F. and Ventura, S. J. (2015) 'Inventorying land availability and suitability for community gardens in Madison, Wisconsin', *Cities and the Environment*, vol 8, no 2, p2.

Etshekape, P. G., Atangana, A. R. and Khasa, D. P. (2018) 'Tree planting in urban and peri-urban of Kinshasa: Survey of factors facilitating agroforestry adoption', *Urban Forestry & Urban Greening*, vol 30, pp12–23.

Ferguson, R. S. and Lovell, S. T. (2014) 'Permaculture for agroecology: Design, movement, practice, and worldview. A review', *Agronomy for Sustainable Development*, vol 34, no 2, pp251–274.

Gori, A., Ferrini, F. and Fini, A. (2019) 'Growing healthy food under heavy metal pollution load: Overview and major challenges of tree based edible landscapes', *Urban Forestry & Urban Greening*, vol 38, pp403–406.

Hurley, P. T. and Emery, M. R. (2018) 'Locating provisioning ecosystem services in urban forests: Forageable woody species in New York City, USA', *Landscape and Urban Planning*, vol 170, pp266–275.

Ickowitz, A., Powell, B. Salim, M. A. and Sunderland, T. C. H. (2014) 'Dietary quality and tree cover in Africa', *Global Environmental Change-Human and Policy Dimensions*, vol 24, pp287–294.

Jacke, D. and Toensmeier, E. (2005) *Edible Forest Gardens, Volume Two: Design and Practice*. White River Junction, VT: Chelsea Green.

Kowalski, J. M. and Conway, T. M. (2019) 'Branching out: The inclusion of urban food trees in Canadian urban forest management plans', *Urban Forestry & Urban Greening*, vol 45, p126142.

Kuo, F. E., Bacaicoa, M. and Sullivan, W. C. (1998) 'Transforming inner-city landscapes - Trees, sense of safety, and preference', *Environment and Behavior*, vol 30, no 1, pp28–59.

Lafontaine-Messier, M., Gelinas, N. and Olivier, A. (2016) 'Profitability of food trees planted in urban public green areas', *Urban Forestry & Urban Greening*, vol 16, pp197–207.

Landor-Yamagata, J. L., Kowarik, I. and Fischer, L. K. (2018) 'Urban foraging in Berlin: People, plants and practices within the metropolitan green infrastructure', *Sustainability*, vol 10, no 6, p1873. Available at https://doi.org/10.3390/su10061873.

Lehmann, L. M., Lysák, M., Schafer, L. and Henriksen, C. B. (2019) 'Quantification of the understorey contribution to carbon storage in a peri-urban temperate food forest', *Urban Forestry & Urban Greening*, vol 45, p126359.

Lovell, S. T. (2010) 'Multifunctional urban agriculture for sustainable land use planning', *Sustainability*, vol 2, no 8, pp2499–2522.

Lovell, S. T., Dupraz, C., Gold, M., Jose, S., Revord, R., Stanek, E. and Wolz, K. J. (2017) 'Temperate agroforestry research: Considering multifunctional woody polycultures and the design of long-term trials', *Agroforestry Systems*, vol 92, no 5, pp1397–1415.

Lovell, S. T. and Taylor, J. R. (2013) 'Supplying urban ecosystem services through multifunctional green infrastructure', *Ecology & Society*, vol 28, no 8, pp1447–1463.

LVEJO (2019) 'Little Village Environmental Justice Organization', http://www.lvejo.org/about-us/history/. Accessed 18 December 2019.

Lwasa, S., Mugagga, F., Wahab, B., Simon, D., Connors, J. P. and Griffith, C. (2015) 'A meta-analysis of urban and peri-urban agriculture and forestry in mediating climate change', *Current Opinion in Environmental Sustainability*, vol 13, pp68–73.

McLain, R., Poe, M., Hurley, P. T., Lecompte-Mastenbrook, J. and Emery, M. R. (2012) 'Producing edible landscapes in Seattle's urban forest', *Urban Forestry & Urban Greening*, vol 11, no 2, pp187–194.

Miljković, J. M. and Crnčević, T. (2019) 'Multifunctional urban agriculture and agroforestry for sustainable land use planning in the context of climate change in Serbia', in Castro, P., Azul, A., Leal Filho, W. and Azeiteiro, U. (eds), *Climate Change-Resilient Agriculture and Agroforestry* (pp197–213). Cham: Springer.

NOAA (2013) 'National Weather Service Forecast Office', National Weather Services, Accessed 22 February.

Nowak, D. J. and Crane, D. E. (2002) 'Carbon storage and sequestration by urban trees in the USA', *Environmental Pollution*, vol 116, no 3, pp381–389.

O'Brien, E. A. (2006) 'A question of value: What do trees and forests mean to people in Vermont?', *Landscape Research*, vol 31, no 3, pp257–275.

Ordonez, C. and Duinker, P. N. (2013) 'An analysis of urban forest management plans in Canada: Implications for urban forest management', *Landscape and Urban Planning*, vol 116, pp36–47.

Park, H., Kramer, M., Rhemtulla, J. M. and Konijnendijk, C. (2019) 'Urban food systems that involve trees in Northern America and Europe: A scoping review', *Urban Forestry & Urban Greening*, vol 45, p.126360.

Park, H., Turner, N. and Higgs, E. (2018) 'Exploring the potential of food forestry to assist in ecological restoration in North America and beyond', *Restoration Ecology*, vol 26, no 2, pp284–293.

Poe, M. R., McLain, R. J., Emery, M. and Hurley, P. T. (2013) 'Urban forest justice and the rights to wild foods, medicines, and materials in the city', *Human Ecology*, vol 41, no 3, pp409–422.

Reid, R. (2019) 'Australian master tree grower program', http://www.agroforestry.net.au/main.asp?_=australian%20master%20treegrower. Accessed 18 December 2019.

Robinson, C., Cloutier, S. and Eakin, H. (2017) 'Examining the business case and models for sustainable multifunctional edible landscaping enterprises in the Phoenix Metro Area', *Sustainability*, vol 9, no 12, p2307. Available at https://doi.org/10.3390/su9122307.

Salbitano, F., Borelli, S. and Sanesi, G. (2014) 'Urban forestry and agroforestry'. In de Zeeuw, H. and Drechsel, P. (eds), *Cities and Agriculture - Developing Resilient Urban Food Systems* (pp285–311). London: Earthscan/Routledge.

Samsøe-Petersen, L., Larsen, E. H., Larsen, P. B. and Bruun, P. (2002) 'Uptake of trace elements and PAHs by fruit and vegetables from contaminated soils', *Environmental Science & Technology*, vol 36, no 14, pp3057–3063.

Schaffer, C., Eksvärd, K. and Björklund, J. (2019) 'Can agroforestry grow beyond its niche and contribute to a transition towards sustainable agriculture in Sweden?', *Sustainability*, vol 11, no 13, p3522. Available at https://doi.org/10.3390/su11133522.

Shackleton, C. M., Hurley, P. T., Dahlberg, A. C., Emery, M. R. and Nagendra, H. (2017) 'Urban foraging: A ubiquitous human practice overlooked by urban planners, policy, and research', *Sustainability*, vol 9, no 10, p1884. Available at https://doi.org/10.3390/su9101884.

Sullivan, W. C., Anderson, O. M. and Lovell, S. T. (2004) 'Agricultural buffers at the rural-urban fringe: An examination of approval by farmers, residents, and academics in the Midwestern United States', *Landscape and Urban Planning*, vol 69, nos 2–3, pp299–313.

Synk, C. M., Kim, B. F., Davis, C. A., Harding, J., Rogers, V., Hurley, P. T., Emery, M. R. and Nachman, K. E. (2017) 'Gathering Baltimore's bounty: Characterizing behaviors, motivations, and barriers of foragers in an urban ecosystem', *Urban Forestry & Urban Greening*, vol 28, pp97–102.

Taylor, J. R. and Lovell, S. T. (2012) 'Mapping public and private spaces of urban agriculture in Chicago through the analysis of high-resolution aerial images in Google Earth', *Landscape and Urban Planning*, vol 108, no 1, pp57–70.

Taylor, J. R. and Lovell, S. T. (2014) 'Urban home food gardens in the Global North: Research traditions and future directions', *Agriculture and Human Values*, vol 31, no 2, pp285–305.

Taylor, J. R., Lovell, S. T., Wortman, S. E. and Chan, M. (2017) 'Ecosystem services and tradeoffs in the home food gardens of African American, Chinese-origin and Mexican-origin households in Chicago, IL', *Renewable Agriculture and Food Systems*, vol 32, no 1, pp69–86.

University of Illinois Extension (2019a) 'Core program areas', https://extension.illinois.edu/global/core-program-areas. Accessed 18 December 2019.

University of Illinois Extension (2019b) 'Illinois master gardeners', https://web.extension.illinois.edu/mg/. Accessed 18 December 2019.

US Forest Service, USDA. (2016) 'Research & development - Green and resilient cities', https://www.fs.fed.us/research/urban/resilience.php. Accessed 6 November 2019.

Wielemaker, R.C., Weijma, J. and Zeeman, G. (2018) 'Harvest to harvest: Recovering nutrients with new sanitation systems for reuse in urban agriculture', *Resources, Conservation and Recycling*, vol 128, pp426–437.

Wortman, S. E. and Lovell, S. T. (2013) 'Environmental challenges threatening the growth of urban agriculture in the United States', *Journal of Environmental Quality*, vol 42, no 5, pp1283–1294.

6

SOILS, INDUSTRIALISED CITIES AND CONTAMINANTS

Challenges for an agroecological urbanism

Salvatore Engel-Di Mauro

1. Introduction

Everyday life in most industrialised cities seems to preclude much interest in farming or soils, unlike food. For those with sufficient money, food magically appears on shelves, tables and other serving places, ready for the taking. To the many who can barely make ends meet, food, especially nutritious food, can be more difficult to get and may require much more effort than for the well-to-do. For the most part, the ecological and social conditions that enable food to be produced and brought to cities are entirely hidden unless people question how things happen, organise, share experiences and actively seek out the means to produce their own food. There are many historical factors producing an elision – physical as well as ideological – of the main sources of food, and each place has a different story to tell about how this has come about. Generally, the development and worldwide spread of capitalist relationships generated a town and country split, sometimes reaching obsessive proportions with proscriptions against much or any farming within city boundaries. It is difficult to break out of the town-country divide, but the split has not been applied everywhere. There remain many urban settings where agriculture has never been expunged and provides major sources of food, mainly involving women's, largely unpaid, work (Orsini et al., 2103).

A recent resurgence in gardening in the more industrialised countries, largely since the 1970s, should not be confused for global processes or for any prospects of food self-sufficiency. They do, importantly, represent a turn against alienation from the sources of our sustenance and for a different urbanism, one that does not take food production for granted, among other activities and relationships imbricated within an agro-food system. The turn involves facing substantive biophysical and political challenges, as urban cultivators know only too well.

These challenges are numerous and only a few are discussed here, mainly in reference to soils and trace element (TE) contamination, the scourge of every industrialised city. The term industrialised cities here refers not only to urban centres characterised by a current or former preponderance of manufacturing, but also of motorised vehicle transport, paved areas and the concentration of toxic substances (through landfills, office buildings, etc.). Understanding soil and health-related concerns and devising alternatives to meet biophysical challenges requires some discussion of farming (or, rather, cultivation) as well as underlying political issues.

Agroecology is a main approach useful in developing ecologically sound techniques in tune with socially constructive ends, as the editors of this volume emphasise. After briefly describing such an approach, I discuss soils relative to their agroecological importance, uniqueness, general characteristics, formation processes and particularities in urban contexts. This is followed by an introduction to the general challenges urban soils present to food production, with special attention to sources and processes of TE contamination. Lastly, I highlight some of the political implications of elevated TE levels and conclude with some thoughts and recommendations on how agroecological urbanism may help confront these kinds of biophysical and social challenges.

2. Agroecology

In large measure, the increasing sensibility towards food production and farming emerges from wider environmental and social concerns. Among them is an awareness that not all has been well with the kind of agriculture that provides the basis for the very existence of industrialised cities. As Gliessman states:

> To create the food productivity that we take for granted today, the industrial system of food production is sacrificing the basic foundations of agriculture – fertile soil, available moisture, amenable climate, nutrient recycling, genetic diversity, and the ecosystem services of natural systems. (Gliessman, 2015: 3)

Profit-centred, capital-intensive, mechanised, agrochemicals and fossil-fuel dependent farming has resulted in massive yield increases for a few crops (mainly cereals) to levels without historical precedent, particularly since the 1950s. Outcomes worldwide have been as life-enhancing for the many as they have been destructive for the many more. Steep productivity rises, especially since the 1950s, have been accompanied by overproduction problems, environmental devastation and social destabilisation. The political agribusiness and latifundist bias intrinsic to conventional agriculture ensures that major gains mean simultaneously major losses, including ecosystem and human health deterioration for many, losses of land for most farmers and curtailment of food access for hundreds of millions. As many places in the world are overstocked with food partly thrown away, many people in underserved (often resource-exploited) and environmentally degraded locations elsewhere, or even within the same overfed places, suffer from deprivation of basic necessities, including potable water, uncontaminated soil and air, food and land. The historical expansion of conventional farming is also among the major factors, such as financial speculation, contributing to spiralling urban enlargement, as the livelihoods of rural dwellers have been steadily undermined, prompting mass migration to cities. Such tendencies are intimately related to the often violent theft of communal resources and the denial or erosion, depending on context, of most people's control over the means of survival and social reproduction (Mies and Bennholdt-Thomsen, 1999; Holt-Giménez, 2017).

Widespread deep dissatisfaction with conventional farming and its global repercussions since at least the 1960s has resulted in the rise of several movements – including within universities – that look to political alternatives while developing ecologically and socially sensible technical options (Gliessman, 2016). A main and influential variant, especially in the Tropics, has been agroecology, which overlaps with organic farming movements largely emerging within industrialised countries. Decades of research and learning from smallholder farming communities have led to the basic recommendations for the development of agricultural techniques that rest on high biodiversity, preserving or increasing soil organic matter content and enhancing nutrient cycling. These techniques have direct political and ecological ramifications: as they help farmers avoid or minimise tillage, monocultures, agrochemicals (e.g., pesticides, synthetic fertilisers) and large-scale irrigation (especially in drylands), they raise smallholders' autonomy and greatly reduce, if not prevent, environmental degradation (e.g., soil erosion, genetic losses, greenhouse gas emissions, soil and water pollution, habitat destruction) as well as perils to human health within farming communities and beyond. Agroecology, however, does not just call into question industrial techniques, but also the capitalist relations that underpin the maldistribution of cultivable land and food, a primary basis of historical disenfranchisement and poverty for many agrarian communities (Altieri, 2009; Gliessman, 2015).

Nevertheless, like organic or sustainable agriculture, increasing institutional acceptance has coincided with technical and market-promoting perspectives overshadowing politically egalitarian ones (Méndez et al., 2016). Technocratic or conventional understandings of agroecology fail to consider (let alone explain) or provide any viable solutions to structural, long-term inequalities, so they are herein largely excluded. Agroecologists who remain true to egalitarian objectives strive not only to establish methods that optimise (not necessarily maximise) food production in ecologically appropriate ways, but also to promote smallholder autonomy, using technical means to reduce or obviate industrial inputs and political strategies to enhance food system control. This implies securing access to land and food by attending to multiple-scale processes, thereby converging with political ecology (Blaikie, 1985; Altieri, 1987; González de Molina, 2016: 61). To achieve such simultaneously ecological and political aims, scientific research is combined with smallholders' knowledge systems by means of transdisciplinary participatory action approaches (Méndez et al., 2016). Despite its complementarity with urban agri-food movements (food justice, urban community gardening, etc.), agroecology as an explicit approach remains largely peripheral to urban food production and alternative agri-food movements (Fernández et al., 2016; Silva and Pfeiffer, 2016). Cuba is among the few exceptions where agroecological principles are formally integrated into urban food production projects (INIFAT, 2010; Leitgeb et al., 2011). At the same time, many urban gardeners already practice agroecological principles without necessarily realising it (Hursthouse and Leitão, 2016; Reynolds and Cohen, 2016).

3. Soils' characteristics, ecological functions and formation processes

One can start drawing together agroecology and urban food justice issues by focusing on soils, a principal means of producing food: these are among other biophysical processes not touched upon here (Gliessman, 2015). Apart from being among the essential requirements of (mainly) terrestrial plants, soils are crucial to biodiversity (within and above soils), the cycling and storage of water and most elements (including nutrients and carbon) and the mitigation of pollution (Hamon et al., 2006). Even in cities where farming has long been banished (and in many is now making a comeback), soils have received more interest of late in industrialised countries, in part due to the spread of concerns over agro-food systems (e.g., localism, organic techniques, community-oriented gardening) and green space (Egorov et al., 2016; Byrne, 2018). Soils play a pivotal role in many aspects of city life. Beyond food production, soils affect the viability of green spaces and buildings, the foundations of transport corridors and flooding levels (Hazelton and Murphy, 2011).

Several processes make soils unique. One is that they are quintessential interfaces of air, water, mineral and organic matter, and life forms. Soil is an evolving set of relationships among organisms, mineral and organic materials, water and air. Soils teem with tiny organisms inhabiting an environment composed of minerals and organic matter of various types and sizes, with spaces in between (pores) that can host varying amounts of water and air. Another principal characteristic of soils is that they are made of layers (called horizons) that form in relation to each other over time. Typically interconnected and interdependent, horizons can be colourfully distinctive, but often they are not easy to discern without some attentiveness to specific physical, chemical and ecological traits. Finally, soils feature an intimate relationship between the mineral and organic (sometimes referred to as clay-humus complexes). Many different kinds of microscopic inorganic and organic substances are bound together, including with microorganisms (Gobat et al., 2004).

All sorts of substances, from plant nutrients to organic and inorganic contaminants, are stored and released through soils. Beyond the manifold support soils provide to plants, many organisms find their main domicile within soils for at least part of their life cycles. The mineral part can be made of broken down (weathered) rocks or loose or compact sediment, which is eventually what becomes of weathered rocks. Soils contain air that is typically heavy in carbon dioxide content, as well as water that can be attached to miniscule particles for decades. What is more, there are often millions of organisms of many sizes and representing thousands of species. Soils can be much more biodiverse than what lies above them. In fact, the life of above-ground plants hinges on soil biodiversity and on the many mutualistic relationships between plants and other organisms like bacteria and fungi. The organisms that dwell in soils are constantly exchanging energy and myriad elements as well as breaking down and storing all sorts of substances. The organic matter that forms largely out of microbial species is what gives soils the ability to store a great deal of water and nutrients (as well as contaminants), to provide many tiny and unique habitats and to show one of their more obvious characteristics, such as darker colouration and at times more sponginess towards the surface (Brady and Weil, 2008; Wall et al., 2015).

Forming soil takes time, from decades to thousands of years. There need to be organisms involved (e.g., lichens, bacteria), together with physical forces (e.g., wind, rainfall), in breaking down or dissolving materials like rocks and organisms' residues and bodies, reconstituting such materials into different substances of diverse sizes and properties (e.g., chemically reactive clay-sized particles, finely degraded and persistent organic matter called humus) and even contributing new materials (Schaetzl and Anderson, 2005). In industrialised cities, soils can be rather ancient (as historical legacies) or even a mere decade for thin soils, but their destruction tends to be much more widespread and much speedier, occurring in less than an hour when excavation machines are used.

4. Soils in urban settings

Since they are subjected to similar kinds of impacts, urban soils tend to develop different characteristics from those in other ecosystems, even if the ecological functions of urban soils differ widely because of their great diversity. Soils may allow water to infiltrate and percolate with relative ease or block water flow, depending on the prevailing particle size (soil texture) and even more so the degree of particle aggregation, together with aggregate shape and size (soil structure). In cities, this relationship between water storage and flow and soils becomes more intricate because of the constant mixing, digging and dumping of materials, implying frequent erosion, disappearance or burial of soils, as well as formation of new soils. The parent material may also be composed of construction debris, broken up asphalt, landfill materials and other human-produced substrates for soil formation. However, soils tend to be compacted because of low organic matter levels due to less microbial activity. Particles dislodged by direct raindrop impact, due to sparse to no protective vegetation cover, collect on soil surfaces and block the openings between soil particles. In combination with water-repellent, waxy covers formed by deposited petroleum-based substances, accumulating displaced particles form impervious surface crusts. Soils also tend to be broken up by buildings, walls and other lasting obstacles that impede water and air flow across soils, a process that also contributes to greater flooding potential. Levels of soil pH tend towards the extreme. Building materials (often carbonate-rich) and large amounts of ash, salts, or pollutants liberated by many kinds of urban activity results in often very alkaline conditions. In some cities, the opposite tendency of much lower pH has been noted, related to acidic building materials or substances emitted by local industries. These inputs and extreme pH values make it difficult for many plant and soil-dwelling species to survive because of low-nutrient availability, if not toxic conditions (Lal, 2018).

Among the most destructive impacts on soils is their partial or total sealing (paving over), leading to the reduction of microbial activity. Partial sealing can enhance contaminant concentrations, while complete sealing contributes to flooding potential and suppresses plant life (Charzyński et al., 2018). Finally, soils in industrialised cities, even those with scant histories of manufacturing, tend to have much higher concentrations of persistent toxins, especially TEs, and polycyclic aromatic hydrocarbons, much as can be found in areas impacted by manufacturing, transportation routes for motorised vehicles, and military and mining activities (Alloway, 2013a; Meuser, 2010). There are also tens of thousands of industrially produced substances, more often organic (i.e., carbon-based), that enter soils, as well as water. These compounds, or their degradation by-products, may be toxic in themselves or in combination with other compounds, and they may remain in soils for days to hundreds of years (and accumulate over time) or be leached out into groundwater (Sauerwein, 2011). All this makes for much greater heterogeneity across very short distances of water (and air) flow and storage as well as biological communities within and on soils, compared to soils typically found in the countryside.

Differences are not just due to sometimes thousands of years of intense human impact, but also to other interrelated factors that lead to the creation and development of soils over time, namely, climate, topography, originating materials (parent material) and the activities of other organisms. So, for instance, some urban soils tend to be poorly drained because of their position at the bottom of a slope near a lake, or they are clay-rich on the surface and so restrict water infiltration. In cities constructed over areas with soils high in sulphate and falling water tables (related to excessive water withdrawals and/or low groundwater recharge due to various human impacts), there can be precipitous falls in pH to levels that can even corrode building foundations (Engel-Di Mauro, 2012). Urban soils may even be contaminated in some TEs because they are formed from parent material derived from underlying rock formations (Bourennane et al., 2010).

Profuse erosion of soil and sediment from the constant bustle of construction is behind sometimes excessive water turbidity and sediment influx, such that streams, if not dredged, eventually become shallower. This magnifies the erosive impacts of extractive industries up- and downstream of cities, such as conventional farming, mining and quarrying. As a result, regular rainfall, that in the past posed little threat of flooding, suddenly becomes a looming hazard (Douglas, 2011). Because of the predominance of paved (sealed) areas in industrialised cities, water in various forms (raindrops, river overflow, etc.) tends to travel and accumulate on the surface instead of at least partly infiltrating in soils or sediments. This, coupled with high amounts of groundwater withdrawals, can impair groundwater recharge and lead to permanently lower water tables, a salient problem in dry regions. In coastal areas, these impacts can also result in seawater seeping into groundwater supplies or, under high-sulphur content conditions, lead to extreme soil acidity (Hazelton and Murphy, 2011).

In cities with moderate to high yearly precipitation, or in arid climates with occasional but intense rainfall, there tends to be a higher frequency of floods or even flash floods and much deeper floodwater levels than would occur if urban surfaces were more water-absorptive. Similar problems happen with compacted, clay-rich, or finer-grained (silty) soil or sediment surfaces, often caused by the constant construction related especially to commercial operations and land speculation. In these kinds of situations, surface openings are either too narrow or blocked entirely, so that water infiltration is impeded or too slow compared to the water flowing. There may also be sub-surface layers impeding the downward percolation of infiltrating water. This can result from physical conditions before a city was built, but it can also happen when sediments are dumped on top of compacted, cemented, or otherwise less pervious materials, or where soils form over decades over those kinds of materials. When it rains or when a river overflows its banks because of too much rain upstream, the water may infiltrate soils or sediments and drain but then begins to slow down or gets blocked when meeting the impervious or less porous layers. In this way, as the water continues to infiltrate above, there is water accumulating until no more water can enter. The net result, as in the case of paved surfaces, is a much greater occurrence of flooding with more devastating impacts (Illgen, 2011).

5. Challenges of urban soils and the ecological importance of food production

Producing food in urban settings meets with many biophysical challenges, as adumbrated above. Conditions for plant growth are highly variable. Within an urban garden there can be small areas conducive to most vegetable production adjacent to others that are unsuitable, if not generally toxic. Some patches may feature thriving microbial communities or optimal pH and nutrient levels, and within a few metres may be the opposite. Soils can have very poor water conductivity because of low organic matter content, high clay content, or compaction, or combinations of these and other variables. Soil water storage may be impeded by a predominance of coarse particles (sand, gravel, or larger) or high sodium content combined with clayey material (creating conditions for underground small tunnel formation), among other similarly impacting characteristics. Water tables may be too high for most vegetables to survive or too low, requiring frequent irrigation, particularly in seasonally or entirely dry climates. As a result of urban heat island effects in industrialised cities, soil temperatures tend to be relatively high, or exhibit large swings; they also tend to heat up too quickly because of low soil organic matter levels. There may be more frequent or magnified extreme events like flooding, landslides, or subsidence, depending on the combination of geomorphological and climatic factors. This does not even touch upon the issue of bioavailable contaminants (human-derived or not), which can impair vegetable growth directly or indirectly (by stifling microbial populations). Hence, even when land access or tenure is secured, together with all the means necessary to produce food, one is confronted with land of highly variable quality under threat of pollution or sudden destructive phenomena.

Although quantitatively insufficient in covering nutritional needs (WinklerPrins, 2017), cultivation in industrialised cities can alleviate some of the intense impacts outlined above if put into practice according to agroecological production principles. In large measure, by happy convergence, cultivators already employ such methods without much interaction with agroecologists or soil management experts, as stated above. The tendency is a preference for organic techniques, virtual absence of tillage, avoidance of most agrochemicals, and, in some instances, attentiveness to soil organic matter replenishment. These kinds of practices improve many soil properties helpful in reducing if not preventing the likes of erosion, airborne dust, flooding and soil sealing. This is because permanent and diverse vegetation cover with organic matter additions improves soil drainage, soil biodiversity, water and nutrient flow and storage, and reduces contaminant availability (within soils or by way of detached airborne particles), to name a few benefits (Lal, 2018). Regrettably, recourse is often made to importing sediment ('soil') to overcome contamination or other problems. This contradicts agroecological understandings of enhancing local ecological relationships without harming others elsewhere. This could be avoided by raising production of, and reliance on, locally composted materials and applying chemical containment measures to reduce or prevent contaminant mobility (increasing soil organic matter already works towards such ends, for example).

6. Recalcitrant contaminants in urban soils: processes and effects

While it promotes general ecological benefits in a number of ways, producing food in cities also carries socially differentiated health hazards (affected by who gets what sort of land in which part of town). In this light, it is surprising that environmental justice activism and research have yet to appreciate the linkage between soil contamination and social injustice or racism, with rare but oblique exceptions (McClintock, 2012). The hazard is largely in the potential exposure to contaminants traceable to working urban soils or in eating contaminated produce. It must also be borne in mind that this is hardly a problem confined to city-grown food. There are debates on the feasibility of growing safe food in cities and over growers' health risks. Some indicate that food and working conditions can be made safe, with qualifications and exceptions that are specific to site, crop and other site factors. The matter is usually presented as a problem that can be overcome technically, using imported soil, avoiding toxin-accumulating crops, and other such prophylactic measures (Clark et al., 2008; Knapp et al, 2016). Others suggest the problem is much more difficult to avoid, especially in poorer, i.e., most, countries, where wastewater use often introduces trace elements (TEs), organic pollutants and pathogens (Clark et al., 2008; Agbenin et al., 2010). Some also question whether the contaminant exposure hazard is any different if not lower than other public health hazards common to cities, like air pollution from vehicular traffic and industrialised settings more broadly (Brown et al., 2016; Hursthouse and Leitão, 2016: 142).

High TE concentrations comprise among the most dangerous, and arguably most intractable, contamination problem, including heavy metals and metalloids (e.g., arsenic). In themselves, TEs are not necessarily problematic; in fact, many of them are important micronutrients for many organisms, including humans. Copper, iron and zinc, for example, are essential for both vascular plants and vertebrate animals. Even chromium (trivalent), molybdenum, selenium and vanadium are important to our health. The issue relates to quantity: it is when present in large amounts that TEs become toxic. However, unlike other contaminants, they are a long-term, if not permanent, toxic hazard, once concentrated. They cannot be broken down and made into inert or less toxic substances, like many organic pollutants, and they cannot be easily diluted into the environment. People affected by high TE levels can suffer long-term health impairment or shortened lifespans. Various forms of cancer are traceable to arsenic and nickel toxicity; neural damage occurs with arsenic, mercury and lead contamination; and kidney damage and bone fragility results from high levels of cadmium exposure (Centeno et al., 2005; Alloway, 2013b). These are among the reasons that TEs have been a main focus in assessing the viability of urban food production (Meuser, 2010; Wortman and Lovell, 2013).

With respect to soil dynamics and urban cultivation, principal contaminant pathways to the human body consist of inhalation and ingestion of soil particles, including those lodged in vegetables through splash and local re-deposition, as well as ingestion of contaminated vegetables (Brevik and Burgess, 2012: 73; Brown et al., 2016: 28). Sources of soil TE contamination include legacies from past land use and proximity to polluting sources (e.g., power plants and incinerators) and vehicular traffic (Alloway, 2004), but also pre-existing substrate (geogenic) sources unrelated to human impacts (Alloway, 2013a; Pouyat et al., 2007). TE transfer from soil to vegetables is mainly governed by pH, amounts and type of clay and organic matter, mineralogy, cation exchange capacity and redox (e.g., waterlogged or dry) conditions (Tack, 2010; Young, 2012). Some TEs become soluble (hence plant available) in soil water under low pH (e.g., cadmium), and others become soluble under high pH (e.g., arsenic); however, their solubility only means that they are potentially plant available (i.e., can be absorbed by roots). Some parts of soluble TEs can be taken up or further modified to insoluble forms by microbes. They may get in part tied up by organic matter and clays or other minerals and contaminate vegetables to a marginal extent. These are only some of the complications TEs may undergo even when soluble. There are yet still more factors to be considered that further complicate the story. Gardeners' inputs can introduce contaminants or alter soil properties in ways that favour vegetable TE absorption, depending on plant physiology (Bolan and Duraisamy, 2003; Wortman and Lovell, 2013). Different crops have diverse tolerance levels and bioaccumulation rates. Some crops, for instance in the *Brassica* genus (kale, broccoli, etc.), are hyperaccumulators for some TEs (Bourennane et al., 2010; Meuser, 2010: 67). With so many variables involved, total soil TE content does not necessarily correspond to higher vegetable TE levels (Allen and Janssen, 2006; Warming et al., 2015). In urban gardens, soil-borne vegetable TE contamination may also be infrequent because of typically high soil organic matter (Kingery et al., 2001; Hursthouse and Leitão, 2016;) or neutralising substances from sources such as construction debris (Howard and Olszewska, 2011).

There are ways to mitigate contamination exposure, including removing and replacing entire soils or their most contaminated parts, importing soil periodically to replace soil surface layers, capping contaminated surfaces and introducing new soil on top, building and using raised beds, mulching surfaces to keep particles down, maintaining near-neutral pH levels, incorporating organic matter (e.g., compost), and thoroughly washing vegetables prior to consumption. When soil TE maxima are exceeded, gardeners are advised to avoid planting accumulator species, root vegetables and herbs. Taken together, these techniques can help prevent exposure to local re-deposition, root-attachment of TE-bearing soil particles and major processes of soil TE mobilisation leading to root absorption (e.g., Bellows, 1999; Hettariachchi et al., 2016).

However, the assumption is that soils are the main conduit of TE transfer to vegetables, and that nearby emissions (e.g., vehicular traffic, incinerators) and local gardening inputs are context-contingent variables that may further contribute to greater contamination. However, TE influx also occurs by means of various kinds of atmospheric deposition. These include long-range wind transport (McKenna-Neuman, 2011) and local resuspension of soil particles, which includes particle entrainment as well as splash (Alloway, 2013a, 25; Wortman and Lovell, 2013). Soil particle resuspension is known to occur within urban areas (Laidlaw and Filipelli, 2008), and in a few cases it has been recognised as an important source of TE contamination in urban gardens (Clark et al., 2006; Wiseman et al., 2015). Splash-derived re-deposition is another local re-deposition factor (McBride et al., 2014). Airborne TEs are absorbed by leaves or can get lodged within other vegetable tissues (Xiong et al., 2014). Brown et al. (2016) have called attention to local re-deposition of lead-laden soil particles, such that importing soil and creating raised beds may not suffice in TE exposure mitigation (Clark et al., 2008). The degree of contaminant entry varies according to TE and vegetable type, and contamination by leaf absorption has been understudied relative to roots (Paltseva et al., 2018).

7. Some political implications of soil contamination

These considerations are typically not considered by promoters of urban food production; this includes leftists who justifiably see in urban gardens the possibility for building post-capitalist egalitarian alternatives. Yet without understanding the complexity of biophysical processes that affect the extent to which contaminants are dangerous, concerned movements or activists can underestimate health hazards and thereby, in the long run, undermine entire political projects. The exposure-reduction effectiveness of preventive measures, like importing soil and building raised beds, cannot be regarded as recipes to be applied in any city. It is no simple task to establish the feasibility and political potentials of urban food production. Biophysical contexts, not only social ones, have to be studied carefully so as to come up with ecologically sensible, health-preserving, as well as socially just alternatives.

The complexity of soil contamination is determined by way of combined social and biophysical processes. As such, there are simultaneously political ramifications. There are political uses of instances of contamination and formidable obstacles to raising awareness and confronting contamination problems as a result of highly uneven knowledge bases and technological wherewithal among affected communities and technical experts. There are unappreciated ecological and social consequences of decontamination by displacement. There is the matter of typically unaddressed accountability intrinsic to pollution legacies. These are but some of the salient political implications of soil contamination (a thorough examination would require a book-length manuscript).

A lack of thorough understandings of contamination processes can leave communities vulnerable to political attack. Recent work by Cutts et al. (2017), for instance, recounts how, near Sacramento, California, soil contamination was used in the 2004 local government's destruction of Mandella Community Garden, their justification for the draconian measure being the presence of high amounts of lead. What was removed was not only all that the gardeners had built over several decades, but also between 30 and 120 cm of the soil they had cultivated (Cutts et al., 2017: 14). The community garden was replaced with another one and on virtually the same site, accompanied by an adjacent housing unit. The complex was renamed the Fremont Mews, endowed with 52 garden plots that people can access for a nominal fee. What is of special interest here is that local government and construction businesses appealed to soil lead contamination as a major justification to relocate the original community garden and, after much resistance, to destroy that garden altogether. They managed to succeed, as the authors put it, in 'reframing the garden as dangerous dirt in need of purification' (Cutts et al., 2017: 14).

This brings to light the problem of activists not being in a position to counter institutional discourse about contamination with technical studies of their own and on their own terms. For instance, if activists had been privy to studies on soil lead and contamination prevention measures, they could have stated from the outset that lead can be contained and they would have been able to formulate lead management plans. This could include raising pH levels and maintaining a permanent vegetation and mulch cover to restrain soil particles entrainment into the air (Menefee and Hettariachchi, 2018). Few studies exist regarding urban cultivators' understanding of contamination issues. So far they point to great difficulties in terms of even appreciating the existence of potential hazards, let alone knowledge of such processes or what to do about them, at least among Baltimore community gardeners (Kim et al., 2014). This is despite readily available resources from extension agencies in the United States, where much effort has been expended in developing plain-language explanations and recommendations for effective exposure prevention (for a Baltimore-specific example, see also the outreach work from The Johns Hopkins Center for a Livable Future, 2014). Barriers to urban cultivators' understanding of soil contamination processes point to highly uneven knowledge bases and technological wherewithal (e.g., field equipment, laboratories) in most affected communities. There is an unacknowledged politics of knowledge among even the most well-meaning officials and technical experts about access to information, studies, education and the infrastructure needed to conduct sampling and analysis. These essential ingredients to determining the presence and degree of contamination and implementing appropriate measures are typically denied socially in gendered and racialised ways, since most urban food production is carried out by women and people of colour (Orsini et al., 2013; Reynolds and Cohen, 2016).

Relative to ecological and social displacement, site decontamination increasingly involves the translocation of ecological ills to other communities. Where contaminated sediment is dumped and which places will experience soil and sediment losses seem not to have even entered most activists' minds, or those of concerned researchers. Some have been paying attention and looking into, for example, the transfer of 'fill' (excavated materials including soils) from New York City to unauthorised dumping sites in the Hudson Valley further north (Vargas and Cheng, 2018). The location and extent of the problem still requires much investigation, even if it is well known among gardeners and involved institutional officials that importing new soil implies quarrying sediment elsewhere (e.g., nearby beaches), if not the locally ecologically catastrophic removal of topsoil. In these instances of contaminated materials' transfer, no affected communities are even being informed about such hazards, least of all approached for consent. The matter involves private property and/or technocratic state organs, which excludes public involvement unless legal actions are taken (i.e., when damage is already done). The geographically wider ecological and social repercussions of the spread of urban cultivation, especially in the most capitalised cities of North America and Western Europe, continue to be unaccounted for or ignored.

There are even more profound issues further confounding prospects for accountability, at least as widely understood in bourgeois systems (whether liberal democratic or authoritarian). TE and other persistent contaminants often long surpass the longevity of those most responsible for producing the problem. The issue of responsibility for environmental harm is a most pressing yet hardly resolved political issue, or, arguably, resolvable under current political systems. Leaded petrol and paint, for example, involved multiple businesses, from production to distribution to consumption. Lead remains among the most widespread pollutant in all major cities and beyond (Mitchell et al., 2014; Li et al., 2017) so that most urban cultivation projects must become knowledgeable of lead contamination potential, find ways of testing soils and produce for lead, and learn ways to circumvent contaminant exposure. To date no businesses have been called to task to pay for all the damage, past, present and future. This is despite the fact that alternatives to leaded petrol and paint and their dangerous health consequences have long been well known (Kitman, 2000). If responsibility were seriously considered and legally pursued (which is not really the case under current political conditions), it would be impossible for any such businesses to afford clean-up costs, present and future healthcare bills and any other conventionally understood costs (never mind social harms that cannot be rectified monetarily; in the case of neurological damage by lead, one cannot simply buy a new brain or all the social life experiences that made that brain). Another aspect is the diffuse aspect of lead pollution, which goes beyond the ultimately responsible but as yet unaccountable businesses or governments. This aspect involves construction and associated soil erosion, leading to lead-bearing particles being strewn about and redistributed by water and wind within and beyond cities. Finding responsibility for these kinds of physical and social processes would be most arduous, and likely counterproductive, since it would entail blaming almost every institution in existence. Complicating the matter further is the fact that the most responsible may be deceased.

Yet, even if one were able to find an original and still living culprit, this would not solve the fundamental and essential problem, the propensity for capitalist relationships to generate such long-term devastation (Kovel, 2007). Not even the localised long-lasting pollution effects are resolvable by identifying the culpable, who, in the case of Union Carbide (and in so many other such cases), have been able to evade responsibility altogether. Since the causes are inexorably systemic in nature, a systemic change is necessary, one that moves beyond bourgeois notions of individual culpability to one that recognises the need for substantive social control over the ways the economy is set up and run relative to processes such as land use (Mies and Bennholdt-Thomsen, 1999). Instead of socialising costs or redirecting them to the next land owner, the problem of diffuse or unaccountable causality should be seized as an opportunity to argue for the devolution of urban planning and management to all urban inhabitants, leading to reclaiming entire cities as commons, rather than continuing with the current predominant piecemeal approach to land struggle.

8. Conclusions

The numerous variables involved and highly context-dependent nature of contamination processes and differential contaminant exposure suggest that any sort of agroecological urbanism must take great care in studying biophysical and social processes at once and with commensurate attentiveness. The many ecological advantages of urban cultivation do not necessarily translate into the social priorities represented in agroecology and social justice movements. Most urban cultivation occurs under tenuous tenure arrangements, private management, or commercial orientation. There are still many social positives that make urban cultivation important, such as community building, educational opportunities, sometimes even livelihood improvement (Chrisinger and Golden, 2016; Gonzalez et al., 2016), even if issues of empowerment or substantive political participation may not be addressed at all or structural inequalities may even be exacerbated (Sbicca, 2012; Eizenberg, 2013, 144; Tornaghi, 2014; Ioannou et al., 2016). Currently, there is scarcely an urban growers' movement explicitly vying to topple capitalist relationships, never mind directly challenging the legitimacy of private property in land. An exception is by way of rights to housing and squatters movements, who are often threatened with eviction if not attacked violently by institutional or privately hired goons under the aegis of legality (Aguilera et al., 2018; Martínez López, 2018). In the main, questions of inequality (e.g., deploring food insecurity and food deserts), if at all considered, go far short of investigating structural causes and organising to overcome capitalist relationships (McClintock, 2014; Tornaghi, 2017). This is among the reasons why urban cultivation established via squatters' movements should be understood as one major starting point in the development of an agroecological urbanism (Engel-Di Mauro and Cattaneo, 2014; Mudu and Marini, 2018).

Furthermore, given the above-explained contamination processes, what is involved in soils is biophysically much wider than soils *per se*. When consciousness-raising about soil contamination is truly ecological in character and thereby includes closely linked processes, like soil water flow, soil erosion and consequent contaminant-laden dust in the air and the re-deposition of eroded soil particles, then it will be much more difficult for institutions to compartmentalise issues legitimately or brush ecologically informed social justice movements aside. A focus on soil contamination through an ecological lens facilitates the development of a wider critique of the capitalist city and the foundations for wider alliances for a project aiming to end the activities that continue to pollute air, soil and water together, given their intimate connections.

Therefore, the development of an agroecological urbanism, if it is to be really agroecological, must envisage cities as ecosystems within which food production and all associated social relationships are embedded. The implication is an urbanist approach where ecology is not reduced to dealing with social relationships and where biophysical phenomena are given equal prominence to social ones. The task is enormously work-intensive, but developing politically explicit and purposeful cross-training in the social and biophysical sciences, while seeking out and organising existing cross-trained scholars and activists, will help surmount the many institutional obstacles on the way to a truly agroecological urbanism. In addition, doing so could contribute to overcoming the town-country divide endemic to capitalist relationships by drawing together smallholder peasant sovereignty and urban food justice movements that have so far led parallel political lives. There is much at stake in combining agroecological and urban justice perspectives and actions.

Reclaiming urban land, cultivating it and struggling for local food sovereignty is arguably among the main objectives of an agroecological urbanism. However, such objectives run up against the potential for increasing people's exposure to health-compromising contaminants, for redistributing contaminated foods locally, and/or decontamination by means of displaced ecological (and thereby social) harm. Contamination problems have also been used to destroy urban community gardens and replace them with land uses, including gardening, that are more amenable to rent-seeking firms. Such cases underscore the need to become educated in soil contamination processes. Relatedly, uneven knowledge and technologies to cope with contamination should be an incentive for agroecological urbanists (and allied activists) to avail themselves of existing education institutions (by whatever means necessary). They should thoroughly re-educate themselves so as to be able to socialise the means of decontamination and carry out a long overdue process of serious, enduring public health promotion that is not forthcoming from current capitalist urban (or any other) institutions. Equally imperative is ensuring that decontamination in cities is not predicated on ransacking other places and ruining other people's livelihoods. Importing soil or increasing the use of lime to raise pH levels, for instance, means mining elsewhere and the consequent destruction of other ecosystems and the communities that live in them. A lack of soil analysis or of engagement with associated sciences not only leads to missed political opportunities, but also to the scuppering of chances to build an egalitarian society based on meeting everyone's needs without compromising ecosystems and human health.

References

Agbenin, J. O., Welp, G. and Danko, M. (2010) 'Fractionation and prediction of copper, lead, and zinc uptake by two leaf vegetables from their geochemical fractions in urban garden fields in Northern Nigeria', *Communications in Soil Science and Plant Analysis*, vol 41, no 8, pp1028–1041.

Aguilera, T., Bouillon, F. and Lamotte, M. (2018) 'Politiques de l'expulsion: Acteurs, enjeux, effets', *L'Année Sociologique*, vol 68, no 1, pp11–38.

Allen, H. E. and Janssen, C. R. (2006) 'Incorporating bioavailability into criteria for metals', in I. Twardowska, H. E. Allen and M. A. Häggblom (eds) *Soil and Water Pollution Monitoring, Protection, and Remediation*. Springer, Dordrecht.

Alloway, B. J. (2004) 'Contamination of soils in domestic gardens and allotments: A brief overview', *Land Contamination & Reclamation*, vol 12, no 3, pp179–187.

Alloway, B. J. (2013a) 'Sources of heavy metals and metalloids in soils', in B. J. Alloway (ed.) *Heavy Metals in Soils. Trace Metals and Metalloids in Soils and Their Bioavailability*, 3rd ed. Chapman & Hall, London.

Alloway, B. J. (2013b) 'Heavy metals and metalloids as micronutrients for plants and animals', in B. J. Alloway (ed.) *Heavy Metals in Soils. Trace Metals and Metalloids in Soils and Their Bioavailability*, 3rd ed. Chapman & Hall, London.

Altieri, M. (1987) *Agroecology: The Scientific Basis of Alternative Agriculture*. Westview Press, Boulder.

Altieri, M. (2009) 'Agroecology, small farms and food sovereignity', *Monthly Review*, vol 61, pp102–111.

Bellows, A. C. (1999) 'Urban food, health and the environment: The case of Upper Silesia, Poland', in M. Koc, R. MacRae, L. J. A. Mougeot and J. Welsh (eds) *For Hunger-Proof Cities: Sustainable Urban Systems*. IDRC, Ottawa.

Blaikie, P. (1985) *The Political Economy of Soil Erosion in Developing Countries*. Longman, Essex.

Bolan, N. S. and Duraisamy, V. P. (2003) 'Role of inorganic and organic soil amendments on immobilisation and phytoavailability of heavy metals: A review involving specific case studies', *Australian Journal of Soil Research*, vol 41, no 3, pp533–555.

Bourennane, H., Douay, F., Sterckeman, T., Villanneau, E., Ciesielski, H., King, D. and Baize D. (2010) 'Mapping of anthropogenic trace elements inputs in agricultural topsoil from Northern France using enrichment factors', *Geoderma*, vol 157, nos 3–4, pp165–174.

Brady, N. C. and Weil, R. R. (2008) *The Nature and Properties of Soils*, 14th ed. Prentice Hall, Upper Saddle River.

Brevik, E. C. and Burgess, L. C. (2012) *Soils and Human Health*. CRC Press, Boca Raton.

Brown S. L., Chaney, R. L. and Hettiarachchi, G. M. (2016) 'Lead in urban soils: A real or perceived concern for urban agriculture?', *Journal of Environmental Quality*, vol 45, no 1, pp26–36.

Byrne, J. (2018) 'Urban parks, gardens and greenspace', in R. Holifield, J. Chakraborty and G. Walker (eds) *The Routledge Handbook of Environmental Justice*. Routledge, London.

Centeno, J. A., Mullick, F. G., Ishak, K. G., Franks, T. J., Burke, A. P., Koss, M. N., Perl, D. P., Tchounwou, P. B. and Pestaner, J. P. (2005) 'Environmental pathology', in O. Selenius, B. Alloway, J. A. Centeno, R. B. Finkelman, R. Fuge, U. Lindh and P. Smedley (eds) *Essentials of Medical Geology: Impacts of the Natural Environment on Public Health*. Elsevier, Amsterdam.

Charzyński, P., Hulisz, P., Piotrowska-Długosz, A., Kamiński, D. and Plak A. (2018) 'Sealing effects on properties of urban soils', in R. Lal and B. A. Stewart (eds) *Urban Soils (Advances in Soil Science)*. CRC Press, Boca Raton.

Chrisinger, B. W. and Golden S. (2016) 'Urban agriculture and health. What is known, what is possible?', in J. C. Dawson and A. Morales (eds) *Cities of Farmers. Urban Agricultural Practices and Processes*. University of Iowa Press, Iowa City.

Clark, H. F., Brabander, D. J. and Erdil, R. M. (2006) 'Sources, sinks, and exposure pathways of lead in urban gardens', *Journal of Environmental Quality*, vol 35, no 6, pp2066–2074.

Clark, H. F., Hausladen, D. M. and Brabander, D. J. (2008) 'Urban gardens: Lead exposure, recontamination mechanisms, and implications for remediation design', *Environmental Research*, vol 107, no 3, pp312–319.

Cutts, B. B., London, J. K., Meiners, S., Schwarz, K. and Cadenasso, M. L. (2017) 'Moving dirt: Soil, lead, and the dynamic spatial politics of urban gardening', *Local Environment*, vol 22, no 8, pp998–1018.

Douglas, I. (2011) 'Urban hydrology', in I. Douglas, D. Goode, M. Houck and R. Wang (eds) *The Routledge Handbook of Urban Ecology*. Routledge, London.

Egorov, A. I., Mudu, P., Braubach, M. and Martuzzi, M. (eds) (2016) *Urban Green Spaces and Health. A Review of the Evidence*. WHO Regional Office for Europe, Copenhagen.

Eizenberg, E. (2013) *From the Ground Up. Community Gardens in New York City and the Politics of Spatial Transformation*. Routledge, New York.

Engel-Di Mauro, S. (2012) 'Minding history and world-scale dynamics in hazards research: The making of hazardous soils in The Gambia and Hungary', *Journal of Risk Research*, vol 15, no 10, pp1319–1333.

Engel-Di Mauro, S. and Cattaneo, C. (2014) 'Squats in urban ecosystems: Overcoming the social and ecological catastrophes of the capitalist city', in C. Cattaneo and M. Martínez (eds) *The Squatters' Movement in Europe. Commons and Autonomy as Alternatives to Capitalism*. Pluto Press, London.

Fernández, M., Méndez, V. E., Mares, T. and Schattman, R. (2016) 'Agroecology. Food sovereignty, and urban agriculture in the United States', in V. E. Méndez, C. M. Bacon, R. Cohen and S. R. Gliessman (eds) *Agroecology. A Transdisciplinary, Participatory and Action-Oriented Approach*, CRC Press, Boca Raton.

Gliessman, S. R. (2015) *Agroecology. The Ecology of Sustainable Food Systems*, 3rd ed. CRC Press, Boca Raton.

Gliessman, S. R. (2016) 'Roots of resistance to industrial food systems', in V. E. Méndez, C. M. Bacon, R. Cohen and S. R. Gliessman (eds) *Agroecology. A Transdisciplinary, Participatory and Action-Oriented Approach*, CRC Press, Boca Raton.

Gobat, J.-M., Aragno, M., Matthey, W. (2004) *The Living Soil. Fundamentals of Soil Science and Soil Biology*. Science Publishers, Enfield.

Gonzalez, Y., Potteigger, M., Bellows, A. C., Weissman, E. and Mees, C. (2016) 'A case study: Advancing public health through gardens for healthy communities (GHC) in New York City: The role of anti-obesity objectives in urban agriculture policy', in E. Snyder, K. McIvor and S. Brown (eds) *Sowing Seeds in the City: Human Dimensions*. Springer Netherlands, Dordrecht.

González de Molina, M. (2016) 'Political agroecology. An essential tool to promote agrarian sustainability', in V. E. Méndez, C. M. Bacon, R. Cohen and S. R. Gliessman (eds) *Agroecology. A Transdisciplinary, Participatory and Action-Oriented Approach*, CRC Press, Boca Raton.

Hamon, R., McLaughlin, M. and Lombi, E. (eds) (2006) *Natural Attenuation of Trace Element Availability in Soils*. CRC Press, Boca Raton.

Hazelton, P. and Murphy, B. (2011) *Understanding Soils in Urban Environments*. Earthscan, London.

Hettariachchi, G. M., Attanayake, C. P., Defoe, P. P. and Martin, S. E. (2016) 'Mechanisms to reduce risk potential', in E. Hodges Snyder, K. McIvor and S. Brown (eds) *Sowing Seeds in the City. Human Dimensions*. Volume 2. Springer, Dordrecht.

Holt-Giménez, E. (2017) *A Foodie's Guide to Capitalism: Understanding the Political Economy of What We Eat*. Monthly Review, New York.

Howard, J. L. and Olszewska, D. (2011) 'Pedogenesis, geochemical forms of heavy metals, and artifact weathering in an urban soil chronosequence, Detroit, Michigan', *Environmental Pollution*, vol 159, no 3, pp754–61.

Hursthouse, A. S. and Leitão, T. E. (2016) 'Environmental pressures on and the status of urban allotments', in S. Bell, R. Fox-Kämper, N. Keshavarz, M. Benson, S. Caputo, S.

Noori and A. Voigt (eds) *Urban Allotment Gardens in Europe*. Routledge and Earthscan, London.

Illgen, M. (2011) 'Hydrology of urban environments', in J. Niemelä, J. H. Breuste, G. Guntenspergen, N. E. McIntyre, T. Elmqvist and P. James (eds) *Urban Ecology. Patterns, Processes, and Applications*. Oxford University Press, Oxford.

INIFAT (Instituto de Investigaciones Fundamentales en Agricultura Tropical) (2010) *Manual Técnico para Organopónicos, Huertos Intensivos y Organoponía Semiprotegida*. Septima Edición. INIFAT, Habana.

Ioannou, B., Morán, N., Sondermann, M., Certomà, C. and Hardman, M. (2016) 'Grassroots gardening movements. Towards cooperative forms of green urban development?', in S. Bell, R. Fox-Kämper, N. Keshavarz, M. Benson, S. Caputo, S. Noori and A. Voigt (eds) *Urban Allotment Gardens in Europe*. Routledge and Earthscan, London.

Johns Hopkins Center for a Livable Future (2014) *Soil Safety Resource Guide for Urban Food Growers*, available at https://www.jhsph.edu/research/centers-and-institutes/johns-hopkins-center-for-a-livable-future/_pdf/projects/urban-soil-safety/CLF%20Soil%20Safety%20Guide.pdf

Kim, B. F., Poulsen, M. N., Margulies, J. D., Dix, K. L., Palmer, A. M. and Nachman, K. E. (2014) 'Urban community gardeners' knowledge and perceptions of soil contaminant risks', *PLoS ONE*, vol 9, no 2, e87913.

Kingery, W. L., Simpson, A. J. and Hayes, M. H. B. (2001) 'Chemical structures of soil organic matter and their interactions with heavy metals', in H. M. Selim and D. L. Sparks (eds) *Heavy Metals Release in Soils*. Lewis Publishers, Boca Raton.

Kitman, J. L. (2000) 'The secret history of lead', *The Nation*, vol 26, pp11–44.

Knapp, L., Veen, E., Renting, H., Wiskerke, J. S. and Groot, J. C. (2016) 'Vulnerability analysis of urban agriculture projects: A case study of community and entrepreneurial gardens in the Netherlands and Switzerland', *Urban Agriculture & Regional Food Systems*, vol 1, no 1, pp1–13.

Kovel, J. (2007) *The Enemy of Nature. The End of Capitalism or the End of the World?* Zed Books, London.

Laidlaw, M. A. S. and Filippelli, G. M. (2008) 'Resuspension of urban soils as a persistent source of lead poisoning in children: A review and new directions', *Applied Geochemistry*, vol 23, no 8, pp2021–2039.

Lal, R. (2018) 'Feeding megacities by urban agriculture in the 21st century', in R. Lal and B. A. Stewart (eds) *Urban Soils (Advances in Soil Science)*. CRC Press, Boca Raton.

Leitgeb, F., Funes-Monzote, F. R., Kummer, S. and Vogl, C. R. (2011) 'Contribution of farmers' experiments and innovations to Cuba's agricultural innovation system', *Renewable Agriculture and Food Systems*, vol 26, no 4, pp354–367.

Li, I., Cheng, Z., Paltseva, A., Morin, T., Smith, B. and Shaw, R. (2017) 'Lead in New York City soils', in V. I. Vasenev, E. Dovletyarova, Z. Cheng and R. Valentini (eds) *Megacities 2050: Environmental Consequences of Urbanization: Proceedings of the VI International Conference on Landscape Architecture to Support City Sustainable Development*. Springer Geography, Berlin.

Martínez López, M. A. (ed) (2018). *The Urban Politics of Squatters' Movements*. Palgrave Macmillan, New York.

McBride, M. B., Shayler, H. A., Spliethoff, H. M., Mitchell, R. G., Marquez-Bravo, L. G., Ferenz, G. S., Russell-Anelli, J. M., Casey, L. and Bachman, S. (2014) 'Concentrations of lead, cadmium and barium in urban garden-grown vegetables: The impact of soil variables', *Environmental Pollution*, vol 194, pp254–261.

McClintock, N. (2012) 'Assessing soil lead contamination at multiple scales in Oakland, California: Implications for urban agriculture and environmental justice', *Applied Geography*, vol 35, nos 1–2, pp460–473.

McClintock, N. (2014) 'Radical, reformist, and garden-variety neoliberal: Coming to terms with urban agriculture's contradictions', *Local Environment: The International Journal of Justice and Sustainability*, vol 19, no 2, pp147–171.

McKenna-Neuman, C. L. (2011) 'Dust resuspension and chemical mass transport from soil to the atmosphere', in L. J. Thibodeux and D. Mackay (eds) *Handbook of Chemical Mass Transport in the Environment*. CRC Press, Boca Raton.

Méndez, V. E., Bacon, C. M. and Cohen, R. (2016) 'Introduction: Agroecology as a transdisciplinary, participatory, and action-oriented approach', in V. E. Méndez, C. M. Bacon, R. Cohen and S. R. Gliessman (eds) *Agroecology.A Transdisciplinary, Participatory and Action-Oriented Approach*. CRC Press, Boca Raton.

Menefee, D. S. and Hettariachchi, G. M. (2018) 'Contaminants in urban soils: Bioavailability and transfer', in R. Lal and B. A. Stewart (eds) *Urban Soils (Advances in Soil Science)*. CRC Press, Boca Raton.

Meuser, H. (2010) *Contaminated Urban Soils*. Springer, Dordrecht.

Mies, M. and Bennholdt-Thomsen, V. (1999) *The Subsistence Perspective: Beyond the Globalized Economy*. Zed Books, London.

Mitchell, R. G., Spliethoff, H. M., Ribaudo, L. N., Lopp, D. M., Shayler, H. A., Marquez-Bravo, L. G., Lambert, V. T., Ferenz, G. S., Russell-Anelli, J. M., Stone, E. B. and McBride M. B. (2014) 'Lead (Pb) and other metals in New York City community garden soils: Factors influencing contaminant distributions', *Environmental Pollution*, vol 187, pp162–169.

Mudu, P. and Marini, A. (2018) 'Radical urban horticulture for food autonomy: Beyond the community gardens experience', *Antipode*, vol 50, no 2, pp549–573.

Orsini, F., Kahane, R., Nono-Womdin, R. and Giaquinto, G. (2013) 'Urban agriculture in the developing world: A review', *Agronomy for Sustainable Development*, vol 33, no 4, pp695–720.

Paltseva, A., Cheng, Z., Deeb, M., Groffman, P. M., Shaw, R. K. and Maddaloni, M. (2018) 'Accumulation of arsenic and lead in garden-grown vegetables: Factors and mitigation strategies', *Science of the Total Environment*, vols 640–641, pp273–283.

Pouyat, R. V., Yesilonis, I. D., Russell-Anelli, J. and Neerchal, N. K. (2007) 'Soil chemical and physical properties that differentiate urban land-use and cover types', *Soil Science Society of America Journal*, vol 71, no 3, pp1010–1019.

Reynolds, K. and Cohen, N. (eds) (2016) *Beyond the Kale. Urban Agriculture and Social Justice Activism in New York City*. University of Georgia Press, Athens.

Sauerwein, M. (2011) 'Urban soils—Characterization, pollution, and relevance in urban ecosystems', in J. Niemelä, J. H. Breuste, G. Guntenspergen, N. E. McIntyre, T. Elmqvist and P. James (eds) *Urban Ecology. Patterns, Processes, and Applications*. Oxford University Press, Oxford.

Sbicca, J. (2012) 'Eco-queer movement(s) challenging heteronormative space through (re)imagining nature and food', *European Journal of Ecopsychology*, vol 3, pp33–52.

Schaetzl, R. and Anderson, S. (2005) *Soils. Genesis and Geomorphology*. Cambridge University Press, Cambridge.

Silva, E. and Pfeiffer, A. (2016) 'Agroecology of urban farming', in J. C. Dawson and A. Morales (eds) *Cities of Farmers. Urban Agricultural Practices and Processes*. University of Iowa Press, Iowa City.

Tack, F. M. G. (2010) 'Trace elements: General soil chemistry, principles and processes', in P. Hooda (ed) *Trace Elements in Soils*. Blackwell, Oxford.

Tornaghi, C. (2014) 'Critical geography of urban agriculture', *Progress in Human Geography*, vol 38, no 4, pp551–567.

Tornaghi, C. (2017) 'Urban agriculture in the food-disabling city: (Re)defining urban food justice, reimagining a politics of empowerment', *Antipode*, vol 49, no 3, pp781–801.

Vargas, O. and Cheng, Z. (2018) 'Sustainable management of urban soil and fill material', paper presented at *Federation of New York Solid Waste Associations Solid Waste & Recycling Conference*, Bolton Landing, NY, 21 May 2018, http://nyfederation.org/wp-content/uploads/2018/pdf2018/53%20ParisioS.pdf

Wall, D. H., Nielsen, U. N. and Six J. (2015) 'Soil biodiversity and human health', *Nature*, vol 528, no 7580, pp69–76.

Warming, M., Hansen, M. G., Holm, P. E., Magid, J., Hansen, T. H. and Trapp, S. (2015) 'Does intake of trace elements through urban gardening in Copenhagen pose a risk to human health?' *Environmental Pollution*, vol 202, pp17–23.

WinklerPrins, A. M. G. A. (2017) 'Defining and theorizing global urban agriculture', in A. M. G. A. WinklerPrins (ed) *Global Urban Agriculture: Convergence of Theory and Practice between North and South*. CABI, Boston.

Wiseman, C. L. S., Zereini, F. and Püttmann, W. (2015) 'Metal and metalloid accumulation in cultivated urban soils: A medium-term study of trends in Toronto, Canada', *Science of the Total Environment*, vol 538, pp564–572.

Wortman, S. E. and Lovell, S. T. (2013) 'Environmental challenges threatening the growth of urban agriculture in the United States', *Journal of Environmental Quality*, vol 42, no 5, pp1283–1294.

Xiong, T., Leveque, T., Shahid, M., Foucault, Y., Mombo, S. and Dumat, C. (2014) 'Lead and cadmium phytoavailability and human bioaccessibility for vegetables exposed to soil or atmospheric pollution by process ultrafine particles', *Journal of Environmental Quality*, vol 43, no 5, pp1593–1600.

Young, S. D. (2012) 'Chemistry of heavy metals and metalloids in soils', in B. J. Alloway (ed) *Heavy Metals in Soils. Trace Metals and Metalloids in Soils and Their Bioavailability*, 3rd ed. Chapman & Hall, London.

7

THE POTENTIAL OF BIO-INTENSIVE MARKET GARDENING MODELS FOR A TRANSFORMATIVE URBAN AGRICULTURE

Adapting SPIN Farming to Brussels

Noémie Maughan, Nathalie Pipart, Barbara Van Dyck and Marjolein Visser

1. Introduction

Like many European cities, Brussels is an important spawning ground for diverse urban agricultural initiatives, including professional (peri-)urban market gardens. This recent revival is led by farmers who are often involved in larger social movements that call for the prioritisation of the fulfilment of social and environmental aspirations rather than maximisation of profit (Morel and Léger, 2016). In addition to these ambitions, the new generation of farmers also has to deal with the current and concomitantly changing institutional, political, social and economic conditions while establishing their professional activities. Accordingly, urban farming practices reflect ideals and aspirations as well as strategic choices and compromises (Morel and Léger, 2016).

This chapter shares insights from a reflexive research process about ideal visions, aspirations, choices and compromises made along the way in the establishment of the Brussels-based Cycle Farm. We show why and how three of the chapter's authors, together with two urban farmers, collectively explored these issues, and what can be learned in terms of the potential of bio-intensive market gardening models for a transformative urban agriculture.

1.1. Expectations for the 'sustainable city'

A growing body of literature documents the benefits of urban agriculture (Mougeot, 2000; Tornaghi, 2014), highlighting its potential contribution to fresh vegetable production, green space, job creation, stormwater retention, greenhouse gas mitigation, neighbourhood beautification and community-building (Lerner, 2012; McClintock et al., 2012; Thibert, 2012). Encouraged by these promises, many European cities draft policies to foster urban food production[1]: the Brussels Region launched the *Good Food Strategy* in 2016.

One of the *Good Food Strategy*'s major priorities is to increase local food production in an innovative and ecological way.[2] Production should reach 30% of total fruit and vegetation consumption by 2035. In this strategic document, the Brussels regional government acknowledges the potential of urban agriculture for the city through its contribution to several environmental, social and economic objectives. From the government's perspective, urban agriculture is expected to contribute to job and income generation while also performing the important function of providing food, to foster the quality of life of Brussels' inhabitants, to reduce the city's environmental footprint, and to provide ecosystem services. Without specifying any modality, the *Good Food Strategy* also defines as a goal that, 'in 2020 100% of new professional agricultural production projects will be efficient on an environmental, economic and social level' (Ronsmans, 2015, p.16). The strategy underlines the need for adapted urban planning and financial support for the multiplication of urban agriculture projects, and for the development of innovative business models. The prescription for financial support highlights the critical importance and sensitivity of the first, second and third years of activities, which often require specific support. The document also recognises that 'several years are often needed for a project to really become profitable' (*Ibid.* p. 34).

To cope with low investment capacities, a broad range of urban agricultural project leaders have developed a strategy of reliance upon sporadic support from various public funds, including subsidies for specific investments for start-ups, business incubators, research funds, part-time salaries, etc. (Verdonck et al., 2012). However, all these funding schemes require proof that the applicant will be able to reach financial autonomy in a time frame ranging from one to three years. For instance, in order to compete for modest grants (up to €25,000 to be used in one year) specifically for professional urban agriculture (and linked to the *Good Food Strategy*), candidates are required to provide a three-year financial plan and are expected to reach financial independence when the subsidies end. Other funding schemes, such as business incubators, demand financial independence after 18–24 months of financial support. Notwithstanding the financial pressure, these subsidised urban market gardeners are still expected to embrace a longer-term, 'sustainable' perspective (providing all social and environmental benefits expected from urban agriculture), as well as to prove that their profession has the potential to create new 'green' viable jobs. Under such pressure, many professional urban agriculture efforts end up prioritising market-based approaches to develop financially viable initiatives. In this context, where at least part of the funding is related to pro-growth and market-driven rhetoric on the sustainable city, it becomes crucial to explore how urban agricultural practices can develop and retain their transformative potential. We think that transdisciplinarity and participatory action research on broadening viability can play a part in enabling urban farmers to reflect critically on the compromises they make in reaction to the political and institutional contexts, constraints and/or pressures.

1.2. Public funding for participatory action research

Three authors of this chapter took part in a research project (SPINCOOP), funded for three years by the Brussels Regional Institute for Research and Innovation as part of the Co-Create programme (*Action Co-Create*).[3] This programme was launched in 2015, concurrently to but independent from the *Good Food Strategy*, which was steered by the Brussels Minister of Environment and the Brussels' Environmental Agency.

Action Co-Create gathers research centres, civil associations, cooperatives and others around a common goal: jointly producing knowledge through a process of successive cycles of questioning, reflecting on and acting to enhance the resilience of the Brussels Region. The research programme was created in 2015 after identifying the need for supporting novel research approaches to address societal demands, and work with the challenges raised by current urban environments. It exclusively funds projects based on Participatory Action Research principles (PAR) and is the first of its kind in Brussels and in Belgium[4] (Van Dyck et al., 2018). The first call for projects focused exclusively on PAR projects for a sustainable Brussels food system.

At that time, Cycle Farm, a small Brussels-based market gardening cooperative, inspired by the SPIN Farming model, had just started to grow vegetables on very small areas in the south of Brussels. The founders of Cycle Farm submitted a proposal to *Action Co-Create*. The coordinator of the research programme suggested the farmers join forces with our research groups.

The action research project SPINCOOP received a three-year grant to (1) study the adaptations of the bio-intensive model *SPIN Farming* (Small Plot Intensive Farming) to the Brussels context; and (2) explore the development of a collective producers' organisation: the Cycle Farm Cooperative. The main questions guiding our joint analysis were the following: how can this adaptation be viable, both in an economical and agroecological sense, and to what extent can SPIN Farming bring resilience and job creation potential to the Brussels Capital Region? Cycle Farm was the case study of SPINCOOP, our 'living lab', and its two main farmers and founders were directly involved in the research project. The research team included two academic researchers in agronomy and sociology,[5] a consultant, and the two Cycle Farm urban market gardeners.

1.3. A hesitant start

In this context and within a three-year time frame, the action research project was expected to demonstrate if and how urban agriculture projects, such as Cycle Farm, could contribute to the transition towards sustainable urban food systems and reach financial autonomy. As agroecology researchers with different disciplinary backgrounds, we were both sympathetic to the initiative as well as doubtful about its transformative potential. We perceived (current) bio-intensive models (and SPIN Farming specifically) as being fundamentally in opposition to an agroecological approach in several aspects, as we develop below. Nevertheless, we decided to embark on this project for four main reasons:

(1) We observed that bio-intensive models are frequently referred to by the neo-farmers with whom we work. We could therefore assume that those models strongly influence the neo-farmers' vision and practices and believed that their study could greatly contribute to better understand and support urban agriculture, whether or not we favour those models.
(2) We decided that we could not miss out on the exceptional opportunity offered by this programme that enables the development of co-creative research for sustainability (see Section 2.2.). Earlier experiences made us confident about the potential of transdisciplinary research to support agro-ecological transition. Our approach to PAR is informed by critical theory, in the sense of developing an understanding of 'how power in social, political, cultural and economic contexts informs the ways in which people act' (McIntyre, 2007, p. 3). It is also informed by popular education in the way it seeks to generate knowledge that starts from situating experience-based knowledge in a systemic perspective, and that is aimed at developing strategies of emancipatory change (Fals-Borda and Rahman, 1991, McIntyre, 2007, Kindon et al., 2007). Being deeply embedded in a joint research trajectory, we hoped and expected to work with Cycle Farm's neo-farmers on their values, not only to analyse and better understand their visions and the compromises designing the Cycle Farm initiative day by day, but also to contribute to the construction and broadening of this vision.
(3) The nature of the partnership: The market gardeners were the first initiators of this research and we believe success in PAR depends predominantly on partners' motivation.
(4) The potential of developing new knowledge and know-how on the viability of urban agriculture in an agroecological sense.

2. Bio-intensive models of urban market gardening

2.1. Neo-farmers' ideals

Today's urban market gardeners in Europe are often 'neo-farmers'. They are relatively new to agriculture in the sense that they do not come from farming families and do not have any related knowledge heritage. As demonstrated by Morel and Léger (2016), based on a study in France, neo-farmers tend to prioritise the fulfilment of social and environmental aspirations over profit maximisation. They build complex personal and professional projects that generally embrace multiple dimensions. Morel and Léger's (2016) study also shows that the confrontation of the farmers' ideals with the potentials, constraints and opportunities of their context leads to compromises reflected in the actual farming practices.

In addition, in earlier cooperation with neo-farmers in Belgium, we have observed that they strongly refer to and find inspiration in diverse external sources, probably because of a lack of knowledge heritage or a family-model to follow. We identify three different categories of inspiration feeding their ideals and ambitions:

(1) **Philosophies and concepts** such as *natural farming* (Fukuoka, 1992), *permaculture* (Mollison and Holmgren, 1981), or *agroecology* (see Section 3.2.2); these offer principles and guidelines and foster an holistic approach, rather than promoting concrete practices of (market) gardening (Morel, 2016).
(2) **Peasant movements and networks** such as *biodynamic agriculture, living soil,*[6] or *adapted machinery* promoted by *l'Atelier paysan*[7]; these gather peasant-researchers who experiment directly on the field and share open access knowledge.
(3) **Business models** such as *bio-intensive models* tested by Coleman (1995), Jeavons (2001), Fortier (2014), or Stone (2015), that are ready-to-use, practice-oriented and are supported by precise figures and step-by-step guidelines.

All the neo-farmers we recently worked with (nine co-researchers in Co-Create projects, and 25–30 neo-farmers through visits and sporadic collaborations) refer to one or several sources of inspirations from these three categories. As has been documented for other starting market gardeners (Morel, 2016), we observed how the Cycle Farm's gardeners compensate their lack of inherited knowledge by combining theoretical inputs and inspirations from strategies or models promoted by other farmers. Interestingly, bio-intensive models seem to be the main inspiration for most of them. It is therefore worthwhile to look back at their origin, and their underpinning of economic and socio-ecological values.

2.2. Evolving bio-intensive models

In its early development in the 1970s, bio-intensive agriculture aimed to enable small-scale farms and farmers who lack the resources (or desire) to implement commercial chemical and fossil fuel-based forms of agriculture to significantly increase food production and income, predominantly based on local, renewable resources and decreased capital and energy inputs.

Those early models, inspired by biodynamic agriculture (Steiner, 2013) and 19th-century-French-intensive gardening methods, focused on enhancing 'ecological' energies and the co-production of nature-based and human-based dynamics as an alternative to mechanisation, chemical and fossil fuel inputs (Gliessman, 2006; Visser, 2013). They were based on a broad range of techniques with light equipment: double digging, composting, raised beds, mixed cropping, increased planting densities (Jeavons, 2001), seasonal extension, locally optimised timing of sowing and harvesting, prevention-not-treatment approach to weed control, and new hand tools and machinery for fast, light and frequent cultivation (Coleman, 1995, 2009).

Although little scientific literature exists on the historical evolution of bio-intensive models, especially related to the socio-ecological values underpinning them, we have come to believe that bio-intensive models have reached a turning point during the last 10–15 years of their development, in terms of how they define themselves and their purpose. While earlier models promoted a highly diversified agriculture, giving prominence to living energies (Visser, 2013) and their synergies, and allowing only for a strict minimal resort to fossil fuels and chemical inputs, the later bio-intensive models favour high-value crops and rely increasingly on non-living energies.

2.3. Bio-intensive agriculture and the imperative of financial viability

The business development and industrialisation of the organic food sector has led bio-intensive model's representative Eliot Coleman to emphasise the ways in which small, local food producers need to develop competitive advantages to sustain their businesses. According to Coleman (2009), if at one time just being organic was enough, he now advocates for business growth through improved production and marketing to compete. Beyond the question of who is able to compete on price, Coleman also explores the difference between shallow (substitution) organics and deep (redesign) organics. Mirroring the debate between weak and strong sustainability (Vivien, 2005), this discussion plunges deep into questions of economic systems and lifestyles (Coleman, 2009).

In that vein, Jean-Martin Fortier and his market garden have become internationally known for achieving profitability and productivity using biologically intensive cropping systems. Advocating a small-scale economically viable sustainable agriculture, Fortier's low-tech, high-yield methods of production employed on the micro-farm form the basis of his bestselling book and toolkit (Fortier, 2014). His publications gather not only information on small-scale equipment (including the walk-behind tractor (two-wheel or walking tractor)), but also marketing and market garden design. Efficiency of time and space is achieved through multiple successions of plantings, optimising labour and materials and standardisation of the size of growing spaces (facilitating crop rotation, production planning, calculation of soil amendments and use of materials such as irrigation piping and plastic covers).

2.4. SPIN Farming as a newborn bio-intensive models

The SPIN Farming philosophy,[8] as one of the newborn bio-intensive models, finds its origins in the United States and intends to turn backyards, front lawns, or neighbourhood lots into significant suppliers of commercial-grade crops, with 'new levels of productivity and profitability that go far beyond traditional [self-sufficiency] home gardening practices' (Satzewich and Christensen, 2008). Under the initial headline of 'making agriculture accessible to anyone, anywhere' (Satzewich and Christensen, 2008), SPIN Farming has become a successful moneymaking farm business for some of its practitioners. Curtis Stone and his Green Acres Farm (Kelowna)[9] represent a pioneering example of a profitable SPIN Farming project, highly prolific on social media, selling a manual book and a rather expensive online course.

Newborn bio-intensive models such as SPIN Farming were designed in North America, in the middle of an economic slowdown, characterised by dramatic and sudden increases in unemployment. This context might at least partly explain why the discourses of the promoters of the models emerging from that period strongly focus on the models' for financial viability. Fortier et al. (2014), Stone (2015) and Hartman (2015) find a purpose in demonstrating that small-scale farming can be an attractive career change or option for (young) people.

3. Co-researching bio-intensive models

3.1. Defining agroecological principles in the SPIN Farming model

One of the first steps in the SPINCOOP research project was to define the notion of 'viability' in the context of Brussels' urban agriculture projects such as Cycle Farm (Figure 7.1). In the context of SPINCOOP, and after many discussions between the researchers and farmers, we defined the core of 'viability' to be:

- multidimensional: viability cannot and should not be reduced to a question of financial profitability/net profit margins. To adequately evaluate the resilience and solidity of a market gardening project takes a multitude of other parameters such as revenue, workload, well-being, etc.;
- highly contextual and dependent on the vision of the neo-farmers over their project, influenced by personal values and inspirations (see Section 2.1);
- in constant evolution, together with the project-and-neo-farmer's lives, which calls for a long(er)-term monitoring.

Given the lack of methodologies to underpin the viability of professional urban farming, the development of tools and/or indicators (relevant for both the practitioners and researchers) became an important objective of the research project. While trying to grasp the multiple dimensions of the farmers' aspirations in their search for creating a viable farm, we noticed that they could easily state some of their strategic goals, such as desired monthly income, commercial strategy, work load, etc. On the other hand, after many interactions, discussions and interviews with several market gardeners from Cycle Farm and other farms in Brussels and across Belgium,[10] we observed that most of them had great difficulty in expressing the specifics of their social and ecological aspirations or goals.

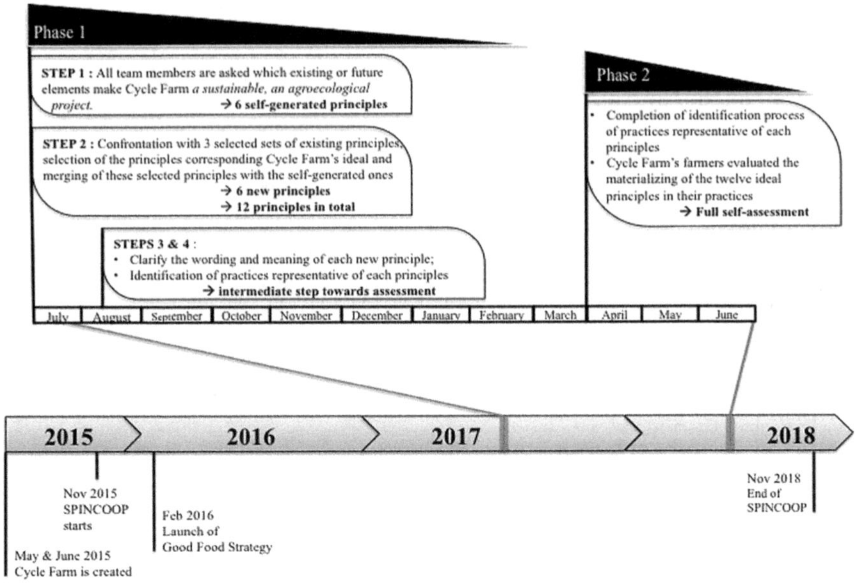

FIGURE 7.1 Time line of the reflexive process on Cycle Farm's ideal, within the SPINCOOP project. *Source:* Authors' elaboration.

This observation was the starting point of our search for new ways to jointly engage in a deeper reflection on the socio-ecological dimensions of urban agriculture. We started by creating some moments explicitly dedicated to reflecting on the visions, values and principles driving Cycle Farm. What started as one initial workshop slowly became a series of workshops, through which we intended to build a common understanding of Cycle Farm's and the two farmers' ecological and social aspirations, their 'agroecological vision' and the influence of SPIN Farming on their ideal. In a second stage, we extended the process to develop a common understanding of the gap between the farmers' 'agroecological vision' and its partial realisation in day-to-day practices (Section 4). This allowed us to clarify the relationship between the imagined ideal, the compromises made in practice, and insights in when, how and why these were made.

3.2. Co-defining an analytical framework: characterising agroecology in Cycle Farm

3.2.1. STEP 1: Auto-generated principles

A series of workshops was initiated by the researchers, then collectively designed and carried out by the entire team. To come to a shared understanding, all team members (both farmers and researchers) were first asked individually to name

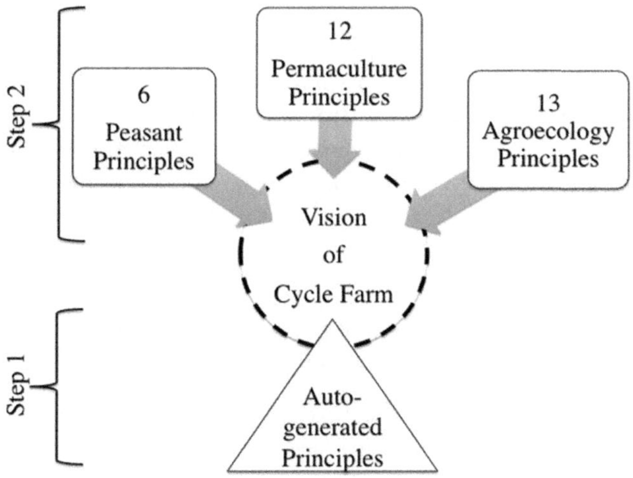

FIGURE 7.2 The two first steps of the reflexive process on Cycle Farm's ideal vision. *Source:* Authors' elaboration.

and describe the existing or future elements that make Cycle Farm a sustainable or agroecological project (Figure 7.2).

These elements were grouped and summarised into 'themes', which were then organised to design a participatory evaluation tool or *Socratic Wheel* (Chevalier and Buckles, 2013). This simple tool has been chosen because it allows the integration of both quantitative and qualitative information, while enabling the participants to be involved in assessing and contextualising the findings as they are created (Schaeffer, 2013).

This first step brought up six auto-generated principles for Cycle Farm. These were described as:

(1) **Maintain a living soil**: Cycle Farm considers the soil as an ecosystem that needs to be 'put back into life' and 'maintained alive' with the least amount of mechanical disruption[11];
(2) **Relocate fertility flows**: Cycle Farm aims to get most of its fertility inputs – organic fertilisers, manure, leaves, etc. – locally and to tend towards circularity insofar as possible;
(3) **Lower carbon footprint**: Cycle Farm focuses on maintaining a low impact on the environment through lesser reliance on fossil fuels regarding mechanisation, inputs and transport;
(4) **Boost functional biodiversity**: Cycle Farm preserves and integrates the spontaneous biodiversity on each cultivated site and boosts biological synergies as key to ecological processes and services;

(5) **Take care of oneself** (invest for comfort/well-being in the long run): Cycle Farm realises that investing in their social and ecological capital in relation with work (in)security, autonomy, intrinsic benefits of work and work-related (dis)comfort is key to secure comfort and well-being in the long run;

(6) **Find a balance between financial viability and feeding people**: Cycle Farm is in search of its own balance between market-oriented strategies, in order to rapidly achieve financial viability, and the aspiration of producing a wide variety of all sorts of vegetables for Brussels' inhabitants at an affordable price.

3.2.2. STEP 2: Confrontation with documented peasant, agroecology and permaculture principles

In a subsequent step, we confronted the first auto-generated thematic clustering with three selected sets of existing principles that have been informing either our work as researchers for a number of years or Cycle Farm's practices. The works on *peasantries* by Van der Ploeg (2008) and *agroecology* as defined by Stassart et al. (2013) have been crucial in shaping our former research in rural contexts. Holmgren (2002) on *permaculture* was added, as this was closer to the imaginaries of Cycle Farm's urban market gardeners (see Table 7.1).

Although the word *permaculture* originally referred to 'permanent agriculture', it was later expanded to 12 agricultural and social design principles centred around simulating or directly utilising the patterns and features observed in natural ecosystems (Holmgren, 2002). Those principles are short, broad and action-oriented.

Stemming from Van der Ploeg's work on the peasant condition and the ongoing re-peasantisation process (Van der Ploeg, 2008), the six peasant principles are associated with the search for greater autonomy and higher performances of the peasant way of farming compared to those of corporate agriculture in terms of efficiency of resource use (water, land, labour, capital), of quality of products, life quality and social inclusion. They rely on strategies to acquire autonomy, such as the creation, development and strengthening of a self-controlled resource base that embraces both social and natural resources, cooperation and reciprocity, resource sharing and high productivity per unit land area through labour-based intensification.

The list of agroecological principles (Stassart et al., 2013) summarised and enriched by the Belgian research group GIRAF[12] assembles *historical* principles developed by Altieri (1995), *methodological* principles and *socio-economic* principles. They range from more 'traditional' agroecological elements, such as soil conditions, agrobiodiversity and biological synergies, to participatory research processes, knowledge and collective adaptation, capacity building, etc.

When confronted with these sets of existing principles as selected by the researchers, Cycle Farm's market gardeners quickly pinpointed principles that would fit with and complete what they had already spontaneously identified.

This second step therefore enabled us to develop, modify and sharpen the auto-generated list and extend it with the following next six principles (Figure 7.3):

(7) **Make better use of water and solar energy**: In order to increase their autonomy and resilience, Cycle Farm aims to limit the use of tap water by collecting and using rainwater. Cycle Farm also relies on the combination of solar energy and selected plant varieties in order to fertilise/capture nutrients in the soil;
(8) **Adapt to micro-heterogeneity**: Cycle Farm observes and interacts with the spatio-temporal variability of resources, adapts to the different microclimates, soil heterogeneities, spontaneous vegetation in place, etc.;
(9) **Feed the local economy**: Cycle Farm reflects on distancing from global markets through reducing the scope of (a) sales, producing locally and directed to the city of Brussels and its hinterland and (b) sources of supply, supporting more local and self-governed providers;
(10) **Put some cleverness back into agriculture:** This rather vast principle could be understood as to 'Learn again to know what, how and why we do what we do'. The idea of 'cleverness' refers to redeveloping know-how through the knowledge of agroecological processes;

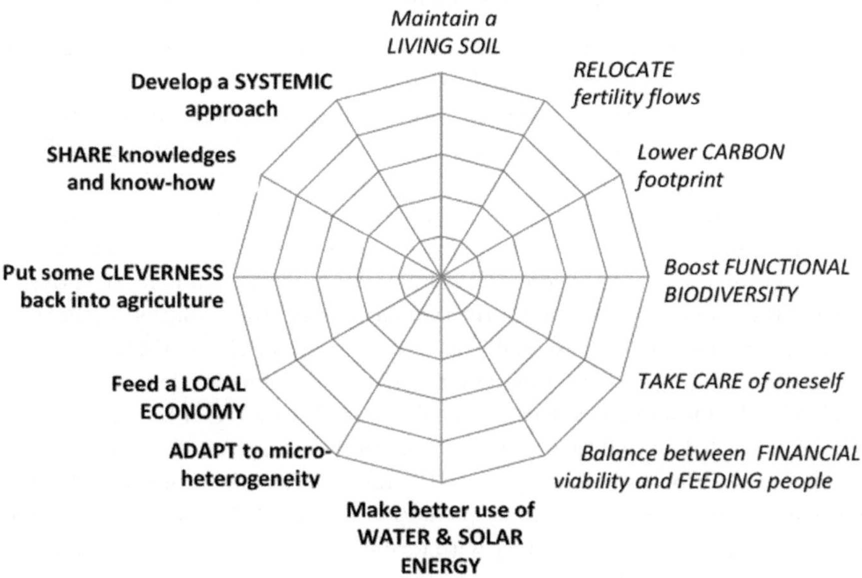

FIGURE 7.3 Cycle Farm's completed set of ideal principles, represented on a Socratic Wheel (Chevalier and Buckles, 2013). Legend: in italic, the auto-generated principles; in **bold** the principles added during phase 2. *Source:* Authors' elaboration.

(11) **Share knowledge and know-how:** Cycle Farm's market gardeners realised that, to nurture their own model of thought, they were sharing (new) knowledge between themselves and with others, including the SPINCOOP research partners, and that they were also brought to transfer these knowledge and know-how to others (trainees, etc.);

(12) **Develop a systemic approach**: Cycle Farm adopts a holistic perspective acknowledging that all elements relate to one another.

Table 7.1 shows the links between Cycle Farm's principles (both auto-generated in the first step and inspirations during the second step) and the three sets of existing principles.

Following this process, during two subsequent meetings (steps 3 and 4 in Figure 7.1), the team went back to the list of principles and clarified the wording and meaning of each new principle. Next, the team identified a few practices for each principle (i.e., ways of working, strategies and choices regarding each aspect of the work) that they perceived as representative of the way the farmers materialise the principles through their everyday work. This intermediate step served as the basis to prepare a full self-assessment during which the farmers evaluated the materialising of the 12 ideal principles in Cycle Farm practices (phase 2 in Figure 7.1).

4. Adaptation of the SPIN Farming model by Cycle Farm

The main inspiration of Cycle Farm's project was officially the SPIN Farming business model. The whole research project was built around its adaptation by Cycle Farm to the Brussels context. In this perspective, we had identified the four key features of the SPIN Farming model that Cycle Farm had explicitly adopted and adapted (see Figure 7.4). Our work in SPINCOOP also focused on a fifth characteristic that Cycle Farm had added to their project in its early conceptualisation, although it was not in the SPIN Farming model: the creation of a cooperative for food production. Although not developed here, the governance of the cooperative by and for the two farmers and the evolution of the work organisation has been a considerable part of the work of SPINCOOP.

As the reflexive process exposed in the previous section revealed, when looked at closely, the ideal of Cycle Farm looks like a patchwork rather than being made from one single fabric. It incorporates inspirations from agroecology, *permaculture* practices (Mollison and Holmgren, 1981), and even further according to the two farmers, inspirations from *natural farming* (Fukuoka, 1992) and *living soil*[13] practices, among others. Concurrently with formulating Cycle Farm's principles, the process allowed us to look deeper into the main features Cycle Farm had taken up from the SPIN Farming model (Figure 7.4), why and how. It revealed the fundamentally hybrid nature of urban agricultural models such as Cycle Farm's, as will be developed hereunder.

TABLE 7.1 Cycle Farm's principles and related inspirations from existing lists of principles

Cycle Farm's principles	*Agroecology list's principles (Stassart et al., 2013)*	*Permaculture principles (Holmgren, 2002)*	*Van Der Ploeg's principles (Van der Ploeg, 2008)*
1. *Maintain a living soil*	Guarantee good soil conditions		
2. *Relocalise fertility flow*	Recycle biomass optimising availability, and balance the flow of nutrients	Use and value renewable resources and services Produce no waste	
3. *Lower carbon footprint*			
4. *Boost functional biodiversity*	Boost diversification (of species and genetic) Boost biological synergies (key ecological processes and services)	Use and value diversity Use edges and value the marginal	
5. *Take care of oneself*			Co-production; labour-based intensification; pluri-activity
6. *Find a balance between financial viability and feeding people*	Explore situations far from local optima already improved	Obtain a yield	
7. Make better use of water and solar energy	Lower the losses of resources Collect water, cover the soil Spatio-temporal variability of resources	Catch and store energy Creatively use and respond to change	
8. Adapt to micro-heterogeneity		Observe and interact	
9. Feed the local economy	Favour autonomy from global markets		Distance oneself from the market (inputs and product (re) produced on site) Social richness

(Continued)

TABLE 7.1 Continued

Cycle Farm's principles	*Agroecology list's principles (Stassart et al., 2013)*	*Permaculture principles (Holmgren, 2002)*	*Van Der Ploeg's principles (Van der Ploeg, 2008)*
10. Put some cleverness back into agriculture	Participatory research Processes Build knowledge and collective adaptation capacities Knowledge diversity		
11. Share knowledge and know-how	Participatory research processes Build knowledge and collective adaptation capacities Knowledge diversity		
12. Develop a systemic approach			

Legend: *Italic*: spontaneous list – first step; Bold: principles added during the second step!

Source: Authors' elaboration.

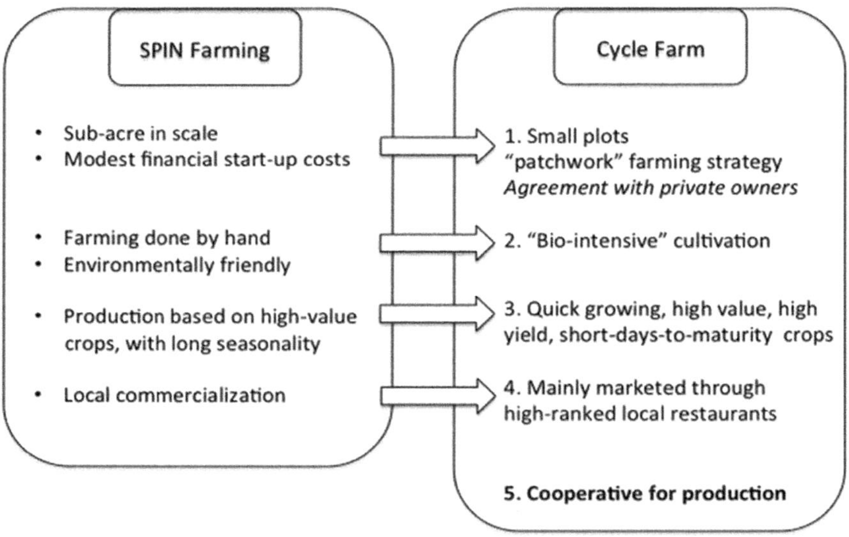

FIGURE 7.4 Adopted and adapted key features between SPIN Farming and Cycle Farm. *Source:* Authors' elaboration.

4.1. Access to land through multiple small plots

In their adaptation of the SPIN Farming model, Cycle Farm's market gardeners chose to access land through negotiation and one-to-one agreements with private landowners to cultivate portions of gardens. In the spring of 2019, they therefore managed five plots that, together, made up approximately 2,400 m^2 of cultivated land (= 0.6 acre).

This small-plot 'patchwork' farming strategy (Figure 7.4), mainstreamed by Curtis Stone, was successful for Cycle Farm: it allowed them to have access to land from their establishment, while gradually increasing their surface area of cultivated land. At the beginning of the third season, Cycle Farm had access to enough land to be able to continue growing activities exclusively on the larger, best located and most productive plots.

However, the patchwork approach also pushes the market gardeners into compromising on several aspects. First, as landowners usually live very close to the plots, negotiations with them sometimes lead to the adaptation of techniques or use of materials such as greenhouses or tarps (items that private landowners preferred not to see in their yard for aesthetic reasons). Second, the agreements between the gardeners and the landowners, referred to as *memorandum of understanding,* are based on precarious contracts. The landowners provide the gardeners with a portion of their land or yard for free or in exchange for small services (weekly vegetable baskets, maintenance work, etc.). The terms of the contract stipulate a preferred time frame for access to the plots (Cycle Farm asks for at least a four-year period), and landowners agree to some basic requirements in case of termination of the contract (such as waiting until the end of the growing season). However, the agreements are not legally binding and can be ended at any time. This absence of long-term secure access to land represents a considerable obstacle to adopting some basic agroecological practices, such as long-term crop rotation, and other practices aimed at restoring and maintaining soil quality. Finally, farming on multiple plots has affected the way in which Cycle Farm's gardeners had initially thought to organise their work, changing from sharing the whole productive process to dividing plots and types of production between them.

Importantly, Cycle Farm differs from Curtis Stone's SPIN Farming approach (that applies the same growing techniques to each plot) and aims to 'Adapt to micro-heterogeneity' (Principle 8), observing and interacting with the spatio-temporal variability of resources, adapting to the different microclimates, soil heterogeneities, spontaneous vegetation in place, etc. 'Maintaining a living soil' (Principle 1) was indeed one of the first principles the Cycle Farm market gardeners spontaneously added to their vision. Through the establishment of permanent beds, organic mulching and working with green manure, Cycle Farm translates this principle into practice.

As land opportunities multiplied, Cycle Farm tended to let go of the smallest and most remote plots and choose larger plots with longer-term access (the access to the largest land that Cycle Farm cultivates is secured through a formal contract, called 'bail à ferme', which includes a binding nine-year minimum provision). The constraints to long-term practices have decreased and they have welcomed other producers on portions of their land (edible flowers and local seed multiplication), creating more diversity.

4.2. Balance between financial viability and feeding people

In SPIN Farming, production focuses on a few high-value crops. Similarly, Cycle Farm focuses on (i) quick-growing, high-value, high-yield popular crops and (ii) marketing through high-ranked local restaurants (Figure 7.4). Cycle Farm production choices, closely linked with this commercial strategy, carry a tension between the mission to 'feed locally', producing a wide variety of all sorts of vegetables at an affordable price, and an attraction to high-value production and high-ranked restaurants in order to improve financial viability. This is reflected, for example, in the choice of not including the agroecology principle 'Valuation of agro-biodiversity as the entry point for re-conceiving systems that guarantee autonomy for the farmers and food sovereignty' (Stassart et al., 2013) in the Cycle Farm's principles.

Nevertheless, if there is an agreement among Cycle Farm's gardeners not to seek agrobiodiversity for the sake of it, they maintain a regular and diverse range of vegetables throughout the year. They grow approximately 40 different species: vegetables, fruits, edible flowers and aromatic herbs.

At the end of the SPINCOOP project, the two Cycle Farm gardeners have expressed diverging opinions on the choices they would like to make to find this balance between financial viability and feeding people. As they had already adapted their work organisation before and shared plots and types of production, they have also multiplied their commercialisation channels: one of them now sells weekly family vegetable baskets in addition to the other distribution channel of the cooperative.

4.3. Local marketing

Principle 9 'Feed the local economy' of Cycle Farm's vision aims to reduce the scope of both sales (producing locally and directed to the city of Brussels and its hinterland) and sources of supply (supporting more local and self-governed providers of seeds, tools, etc.).

Consistent with this principle, local marketing has been predominant in Cycle Farm activities from the outset (Figure 7.4). Similar to Stone's SPIN Farming model, Cycle Farm started their commercial activity early on by soliciting and building commercial relationships with local (high-ranked) restaurants. This has gradually expanded to a local market and organic stores in the vicinity.

4.4. On seed autonomy

On the particular issue of seed autonomy, we have observed a shift in Cycle Farm's positioning throughout the process. A few cooperatives exist in Belgium and the rest of Europe, such as *Bingenheimer*[14] (Germany), the more recent *Cycle en Terre*[15] (Belgium), or the network *Zelf Zaden Telen*[16] (Belgium). These as well as other options, such as joining other seed networks or reproducing local seeds on-farm, make relying on productive, diversified and adapted local seeds varieties possible, although still very challenging given the specific surface areas required, the technical nature of the process and the high risk of failure.

However, during the preparation of its first production season (2015), the fact that seed suppliers were corporate was not a major decision factor for Cycle Farm. They had taken advantage of an overseas equipment order to have all the seed shipped from the United States. They considered the idea to test traditional/peasant varieties during the first years to be utopian. They attribute this decision to the lack of time and/or of skills and motivation: they do not know how to and do not want to become seed producers. Besides, urban market gardeners taking part in this have to reserve a specific area to multiply the seeds. '*I am not sure it is viable, and my first interest is my viability*' (Cycle Farm, July 2017).

In the first phase, when researchers put forward the possibility of gradually abandoning hybrid varieties and trying more open-pollinated varieties, this suggestion remained unanswered by the farmers. While the Cycle Farm discourse indicated that they were aware of this issue's important and central place in the struggle of many producers to regain autonomy, they clearly stated that it was absolutely not a priority for them at this point.

During the second and third phases of the visioning workshops however, the use of locally produced seeds of open-pollinated varieties reappeared when operationalising the practices and indicators for Principle 9 'Feed the local economy'. They tailored one of the indicators to measure the *localness* of their supply.[17] Cycle Farm has also very recently allowed and welcomed another producer to come and cultivate seeds on small portions of their land.

5. Concluding reflections

Our analysis shows that although Cycle Farm officially gets its main inspiration from SPIN Farming (a bio-intensive business model), the ideals guiding the two farmers of this cooperative are much more complex. Their social and ecological aspirations are diverse; they are not limited to short-term economic profitability and are much more connected to other sources of inspirations than the farmers could spontaneously express: agroecology, permaculture, etc.

However, despite this diversity of aspirations, during the full three years that we cooperated with Cycle Farm, its gardeners have continued to broadly identify their financial viability as the first hurdle to overcome, before being able to explore other dimensions of resilience (Plateau et al., 2019). Following the last generation of bio-intensive models, Cycle Farm has given priority to high-value crops, high-ranked local commercialisation partners and multiple 'disposable' plots. Until now, most of the discussions and decisions outline this opposition between short-term financial/economic viability and long-term viability (social, environmental, political). Even though Cycle Farm is conscious that investment in soil fertility, biodiversity, farmers' well-being, etc. might bring beneficial consequences in the mid- to long-term, the main conscious decision-making factor and priority is the direct and short-term impact on financial viability. The mid- to long-term ecological and social investments that are part of their ideal project are either initiated if they are synergetic or at least not in competition with improvement of financial viability, e.g., practices related to Principles 2 'Relocate fertility flows', 3 'Lower carbon footprint' and 7 'Rely more on water and solar energy', or they were postponed to a hypothetic later when they would become feasible. Cycle Farm's gardeners expressed to have made conscious choices to set aside some of their agroecological aspirations, focusing on the production objective for the first few years.

Several reasons could contribute to explain these strategic choices. The specific context of the start-up phase could be one of them: this phase is characterised by high uncertainties regarding the potential durability of the project, concerns that can last several years before a relative stabilisation. However, this stage also presents a potential for exploration of innovative trajectories and a wide scope of possibilities (Gasselin et al., 2012). The last argument would seem to be even more relevant if the new professional activity were to be financially supported during those first years, enabling them to experiment alternative or '*avant-gardist*' practices, precisely what is expected from the Brussels' farmers at Cycle Farm. However, although Cycle Farm benefitted from several subsidies in addition to the Co-Create financial support, they did not manage to get rid of the overriding pressure to reach financial viability as soon as possible. As our research shows, despite being partly subsidised, the farmers did not feel sufficiently relieved of financial pressure to invest broadly in the long-term agroecological practices that are part of their ideal project. This observation leads us to see a paradox in the current political and institutional context.

The Brussels' regional institutions have launched political strategies and research funding frameworks to explicitly promote innovative food production initiatives. However, the conditions of the financial support that is offered, while noteworthy, do not match the degree of 'innovation' or 'alternativity' that is acknowledged to be needed to considerably move forward in the transition of the current food systems. These political instruments are still imbued with short-term economic profitability imperatives and pro-growth thinking.

McClintock (2014) described this paradox earlier when showing how public policies initially designed with the intention to support and promote innovative sustainable food systems in fact continue to work within the capitalist extractive logic of the current food system (Allen and Guthman, 2006; Alkon and Mares, 2012) through 'promulgation of neoliberal discourses of personal responsibility and market-based solutions' (McClintock, 2014, p. 149). Current financial support instruments in the Brussels Region, through short time frames and requirements to prove rapid financial independence, accentuate the pressure to prioritise economic profitability over long-term social and ecological aspirations. It bears the risk to push neo-farmers to 'self-greenwash' their ideal and/or to resort to market-based economic solutions rather than taking the time to explore and invest in alternatives.

We experienced this paradox in a very particular and personal way. As researchers and partners in a PAR project, reflecting and acting on the development of agroecological practices for urban food production, we became very aware of the expectations from many external actors. We felt caught in the same paradox that we had unveiled. We see there a second intertwined paradox: while financial viability and pressing time constraints related to the Co-Create PAR subsidy probably limited the potential for exploration of more alternatives, it is also this unique research opportunity that enabled us, a team of academics, counsellors and neo-farmers, to carry out a three-year transdisciplinary study to critically reflect and act on the contexts and conditions in which urban farms emerge.

This brings us to our final reflections: what change for the better did we actually foster by engaging with Cycle Farm over these three years, for Cycle Farm as well as for the wider community of practitioners practising urban market gardening? An even more nagging question is: what change for the better did we actually impede by engaging with Cycle Farm?

Local initiatives that try to 'change the (food) system' are utterly vulnerable, since they experiment both socially and technically at the margins of the very systems they are trying to change. How to sufficiently consider these vulnerabilities and the resulting tensions when engaging in joint research initiatives? (Louah et al., 2015; Van Dam et al., 2017; Van Dam and Visser, 2018). In some cases, the research 'add-on' can actually worsen these tensions, instead of genuinely servicing the initiative. From our experience, an awareness of the necessary change of posture that PAR requires, from both researchers and practitioners, is a necessary step in reducing these tensions. Humility and clarity about roles and decision-making procedures is part and parcel of this required change.

Since PAR emphasises processes over results, mostly over longer time frames than project finance allows, it is actually very hard to pin down what precise change PAR fosters or impedes. It is therefore of utmost importance to negotiate, right from the start, what can reasonably be expected and actively aimed for as 'research outcomes' when funding ends. For a 'good' PAR, a 'good' negotiation might then be defined as one that fosters the required change of attitude.

Acknowledgements

We would like to address our heartfelt thanks to David Errera and Nicolas Vlaminck, the two founders and farmers of Cycle Farm, who did us the honour of involving us in this exceptional adventure. Special thanks as well to Jerôme Rassart, from Credal, for embarking on this journey with us, learning to 'play the game' of PAR with an open mind and sharing his grounded knowledge of neo-farmers' challenges and successes. Our thanks go to Kevin Maréchal and

Lou Plateau, for launching SPINCOOP project with us, and to Margaux Denys and Manon Bertha for their contributions to the research project.

The SPINCOOP project was supported by Action Co-Create, Innoviris. The work of the third author was supported by the EU under the Marie Sklodowska-Curie Action, HealingFromEnclosure 707807.

Notes

1 See, for example http://www.milanurbanfoodpolicypact.org/signatory-cities.
2 For more information, please see https://document.environnement.brussels/opac_css/elecfile/Strat_GoodFood_FR.
3 http://www.cocreate.brussels/-SPINCOOP-?lang=fr ; For more details on the 'Co-create' programme, see https://innoviris.brussels/co-creation.
4 For information on the PAR projects and the support structure, see http://www.cocreate.brussels.
5 The first two authors of this paper. Authors three and four had a supportive role in the research but were not involved in its daily practice.
6 See https://www.biowallonie.com/documentations/maraichage-sol-vivant/.
7 See https://www.latelierpaysan.org/.
8 The founder of this 'community' are Roxanne Christensen, President of the *Institute for Innovations in Local Farming* (Philadelphia) and owner of the Somerton Tanks Farm (www.somertontanksfarm.org), and Wally Satzewich and Gail Vandersteen, producers at the Wally's Urban Market Garden, a 'multi-sites' urban farm (Saskatchewan).
9 See http://www.greencityacres.com and http://theurbanfarmer.co.
10 The first two authors of this paper are partners of another Co-create research Ultra-Tree, which involves other other (urban) market gardeners. See https://www.cocreate.brussels/projet/ultratree/.
11 This principle and the practices are inspired by the network of peasants-researchers *Maraichage sur sol vivant*: https://www.facebook.com/maraichagesolvivant/.
12 GIRAF or Groupe Interdisciplinaire belge de Recherche en Agroécologie du FNRS (Belgian Interdisciplinary Agroecology Research Group funded by FNRS) http://www.agroecologie.be.
13 See maraichage sur sol vivant : http://maraichagesolvivant.org.
14 https://www.bingenheimersaatgut.de/.
15 http://www.cycle-en-terre.be/.
16 http://www.zelfzadentelen.be/.
17 On a scale ranging from 0 to 5 (5 being their goal): 0 = Everything bought from outside of Europe; 1 = Mix of from out- and inside Europe; 2 = Mix of from inside Europe and bordering countries; 3 = Bordering countries; 4 = Bordering countries and from companies working with a collective/cooperative mode of governance; 5 = Production of more than 20% of seeds on-farm.

References

Alkon, A. H. and Mares, T. M. (2012): Food sovereignty in US food movements: Radical visions and neo-liberal constraints, *Agriculture and Human Values*, vol 29, no 3, pp.347359.

Allen, P. and Guthman, J. (2006): From "old school" to "farm-to-school": Neoliberalization from the ground up, *Agriculture and Human Values*, vol 23, no 4, pp.401–15.

Altieri, M. A. (1995): *Agroecology: The science of sustainable agriculture*, 2nd edn. Westview Press, Boulder, CO.

Lerner, S. (2012): *Seattle food action plan*. City of Seattle Office of Sustainability and Environment, Seattle.

Louah, L., Visser, M., Baltazar, S. and Delobel, V. (2015): Changements de postures du chercheur, de l'agriculteur et de l'enseignant pour l'innovation agroécologique paysanne, *Pour*, vol 226, pp.5–10.

McClintock, N. (2014): Radical, reformist, and garden-variety neoliberal: Coming to terms with urban agriculture's contradictions, *Local Environment: The International Journal of Justice and Sustainability*, vol 19, no 2, pp.147–171.

McClintock, N., Wooten, H. and Brown, A. H. (2012): Toward a food policy "first step" in Oakland, California: A food policy council's efforts to promote urban agriculture zoning, *Journal of Agriculture, Food Systems, and Community Development*, vol 2, no 4, pp.15–42.

McIntyre, A. (2007): *Participatory action research (PAR)*, Qualitative Research Methods Series, Sage, Thousand Oaks, CA.

Mollison, B. and Holmgren, D. (1981): *Permaculture one: A perennial agriculture for human settlements*. International Tree Crops Institute, Davis, CA.

Morel, K. (2016): *Viabilité des microfermes maraîchères biologiques. Une étude inductive combinant méthodes qualitatives et modélisation*, PhD Thesis, Université Paris-Saclay.

Morel, K. and Léger, F. (2016): A conceptual framework for alternative farmers' strategic choices: The case of French organic market gardening microfarms, *Agroecology and Sustainable Food Systems*, vol 40, no 5, pp.466–492.

Mougeot, L. J. (2000): Urban agriculture: Definition, presence, potentials and risks, and policy challenges, Cities feeding people series; rept. 31.

Plateau, L., Maughan, N., Pipart, N., Visser, M., Hermesse, J. and Maréchal, K. (2019): La viabilité du maraîchage urbain à l'épreuve de l'installation professionnelle, *Cahiers Agricultures*, vol 28, no 6, p.8.

Ronsmans, M. (2015): Good food strategy. Towards a Sustainable Food System in the Brussels-capital region. Available at: https://document.environnement.brussels/opac_css/elecfile/BRO_GoodFood_Strategy_ENGL.pdf [Accessed on March 6, 2020]

Satzewich, W. and Christensen, R. (2008): SPIN farming. Grow food, make money, learn how. Available at: https://spinfarming.com/ [Accessed on 6 March 2020]

Schaeffer, H. (2013): Progress marker indicators for developmental change. Available at: https://www.outcomemapping.ca/download/Progress%20Marker%20Presentation%20Outcome%20Mapping.03.27.13.Final.pdf. [Accessed on 13 August 2019].

Stassart, P. M., Baret, P., Grégoire J. C., Hance, T., Mormont, M., Reheul, D., Stilmant, D., Vanloqueren, G. and Visser, M. (2013): L'agroécologie: Trajectoire et potentiel. Pour une transition vers des systèmes alimentaires durables, in Van Dam, D. (ed.): *Agroécologie: entre pratiques et sciences sociales*. Educagri Editions, Dijon, pp.25–51.

Steiner, R. (2013): *Agriculture course: The birth of the biodynamic method*. Rudolf Steiner Press, Forest Row.

Stone, C. (2015): *The urban farmer: Growing food for profit on leased and borrowed land*. New Society Publishers, Gabriola Island. 288pp.

Thibert, J. (2012): Making local planning work for urban agriculture in the North American context: A view from the ground, *Journal of Planning Education and Research*, vol 32, no 3, pp.349–357.

Tornaghi, C. (2014): Critical geography of urban agriculture, *Progress in Human Geography*, vol 38, no 4, pp.551–567.

Van Dam, D., Lagneaux, S., Nizet, J. and Streith, M. (2017): *Les collectifs en agriculture bio. Entre idéalisation et réalisation*. Educagri Editions, Dijon.
Van Dam, D. and Visser, M. (2018): Recherche et collectifs agro-alimentaires: Vulnérabilité et tensions, *Pour*, vol 234–235, pp.163–170.
Van der Ploeg, J. D. (2008): *The new peasantries: Struggles for autonomy and sustainability in an era of Empire and Globalization*. Sterling, Earthscan, London, 356pp.
Van Dyck, B., Van Keerberghen, A., Massart, E., Maughan, N. and Visser, M. (2018): Institutionalization of participatory food system research: Encouraging reflexivity and collective relational learning. *Agroecologia*, vol 13, no 1, pp.21–32.
Verdonck, M., Taymans, M., Chapelle, G., Dartevelle, G. and Zaoui, C. (2012): *Système d'alimentation durable. Potentiel d'emplois en Région de Bruxelles-Capitale*, Rapport final de la recherche réalisée pour le compte de l'Institut Bruxellois pour la Gestion de l'Environnement. Bruxelles: Centre d'études régionales bruxelloises, Greenloop S.A. et Observatoire bruxellois de l'emploi.
Visser, M. (2013): Revisiting Ester Boserup: The agroecology of agrarian change under population pressure, in van Ypersele, J.-P. and Hudon, M. (eds): *1er Congrès interdisciplinaire du développement durable : quelle transition pour nos sociétés*, ULB, UCL & Wallonie, Namur, pp 141–149.
Vivien, F. D. (2005): *Le développement soutenable*. La Découverte, collection Repères, Paris.

8

THE TRANSFORMATIVE POTENTIAL OF AGROECOLOGICAL FARMERS

An analysis of participatory food system strategies in Nicaragua and England

Elise Wach and Santiago Ripoll

(With contributions from Chris Smaje, Jorge Irán Vásquez Zeledón, Clare Ferguson and Julio Hector Sanchez Gutierrez)

1. Introduction

Now more than ever, evidence concludes that our food systems are failing to adequately nourish our populations and are simultaneously severely degrading the ecosystems on which they depend. While there is widespread agreement that our food systems need to change in order to more adequately feed the human population in ways that are ecologically sustainable (or even regenerative), there is a wide range of perspectives about what can and should be done in response to these failures.

Strategies and approaches based on the concept of agroecology offer the potential to address many of the problems in the mainstream food system. Agroecological approaches have been shown to increase dietary diversity, maintain or increase yields and sustain or even enhance ecosystems (Altieri, 1987; Sevilla Guzman and Woodgate, 2013; Pretty et al., 2006; Gliessman, 2007; Chappell and La Valle, 2009). Whilst some interpretations of agroecology include an explicit focus on redistributing power within food systems (for example, through valuing farmer knowledge and agency and considering the political and economic structures that shape them (see e.g. Mendez et al., 2013, Gonzalez de Molina, 2013), agroecology as a concept can be, and in some cases has been, limited to technical agricultural approaches that omit the social aspects of agriculture and food (Levidow, 2015). However, a leading proponent has argued that "ecological change in agriculture cannot be promoted without comparable changes in the social, political, cultural, and economic arenas that conform and determine agriculture" (Altieri, 2009).

Holt-Giménez and Shattuck's (2011) quadripartite[1] classification of approaches for addressing food system failures (and for their transformation) indicates that agroecology and food sovereignty movements include trends that could be classified as 'progressive' as well as those that would be considered 'radical'. Progressive trends, according to the authors, include developing alternatives to industrial foods that are 'largely within the economic and political frameworks of existing capitalist food systems' (*Ibid.*, 115). Radical trends include a stronger focus on 'entitlements, structural reforms to markets and property regimes, and class-based, redistributive demands for land, water and resources' (*Ibid.,* 115). The authors argue that progressive trends alone would be insufficient to adequately transform the 'practices, rules and institutions determining the world's food systems' in order to address food system failures, particularly in relation to hunger (*Ibid.,* 132), and call for more alliances between progressive and radical trends.

Tilzey (2017) similarly argues that without 'radical' (or counter-hegemonic) approaches, it will be impossible to adequately redress the problems of our existing food systems. Drawing on a 'Political Marxist' analysis, Tilzey details how the capitalist social system at present, which conceptualises the 'market' and the 'economy' as separate from society and ecologies, will only be able to mitigate, but not resolve, the 'disbenefits' of existing food systems in the social (e.g., livelihoods, dignity, agency, health) and ecological (e.g., soil fertility, biodiversity, climate change) realms. In this system, consumer interests (e.g., cheap food) are often perceived as antithetical to producer interests (e.g., high farm gate prices), and producers are driven to compete with one another to maintain their subsistence and social position (*Ibid.*). Further, with the flow of ecological and social (i.e., labour power) resources in the capitalist world system from the 'periphery' to the core (and a growing 'semi-periphery'), it is the subaltern classes in the Global South that are most likely to engage in radical actions to undermine the existing system. It is these people who 'have the least to lose and the most to gain' (see Chapter 10 by Tilzey). Holt-Gimenez and Shattuck also note that strategies in the Global North tend to be 'progressive' rather than 'radical' (2011: 124).

Within the agroecology and food sovereignty movements, there is an emphasis on the rights of producers to shape their own food systems (LVC 2007) given that the perspectives of producers have typically been marginalised in food system decision-making (Pimbert et al., 2002). The food sovereignty movement has spanned the Global North and Global South (Desmarais, 2007) and amalgamated both progressive and radical strategies in what Patel (2011) has identified as its 'big tent' politics. Bernstein (2014), Edelman et al. (2014) and others have questioned the clarity of the food sovereignty movement's aims and the ability of producers to undermine capitalism given the conflicting positionalities and interests within the movement.

Therefore, there are questions about whether producers of varying positionalities would develop strategies that are sufficiently transformative (or 'radical') to address the problems of our existing food systems. It is also relevant to ask whether strategies of farmers in the so-called Global South are similar or contradictory to those of farmers from the so-called Global North.

Building on a participatory farmer-led research initiative involving producers from Nicaragua and England, this chapter seeks to answer the question, 'What types of strategies might be developed by ecological producers in the Global South and North, and are these sufficient for transforming food systems?' Specifically, it details the framings articulated and approaches identified by farmer panels in England and Nicaragua throughout a participatory research process, and provides an analysis of the extent to which these strategies might have the potential for transforming food systems to become more ecologically and socially sustainable, if not regenerative.

2. The initiative

A participatory initiative that took place in 2017–2018 sought to facilitate ecological food producers in both the Global North and the Global South to develop strategies and action plans to change their food systems through deliberative processes. This project was based on the premise that producers possess intimate and invaluable knowledge about the realities of operating within the existing food system. It was also guided by theories of participatory change processes (Reason and Bradbury, 2008) and on the principles of the food sovereignty movement (LVC, 2007); these indicate that producers need to be involved in, or leading, food system changes.

Through participatory systemic action research (Burns, 2012; Burns and Worsley, 2015), panels of farmers in Nicaragua, Senegal and England, who self-identified as practising ecological farming, sought to identify potential pathways for realising agroecological food systems. The conceptualisations of 'agroecological food systems' themselves were determined by each of the farmer panels, who were encouraged to reflect on their own experiences as farmers and consumers, and to explicitly consider the ecological, nutritional (i.e., public health), social (i.e., community building) and economic (i.e., livelihoods) dimensions of food systems. The research process included participatory complex systems mapping (*Ibid.*), participatory development of research questions, collaborative research, deliberative processes for participatory analysis of research findings (Carpini et al., 2004), identification of leverage points for change based on systems thinking (Senge, 1990; Meadows, 1997, 2008) and the co-development of strategies with potential allies and agents of change identified by the farmer panels in each country. Farmers primarily focused on their own food systems throughout the two years, with consideration of linkages with other countries. Farmers from the three countries gathered together at the end of the project to exchange experiences and insights. The methodology enabled a broadening of discussions to go beyond their immediate concerns to a consideration of wider systemic issues. The participatory process had a strong reflexive component with the deliberations particularly focused on challenging preconceptions.

The composition of the farmer panels in our research makes their insights particularly distinctive: rather than being a representative sample of the farming population in each country, panel members were producers involved directly or indirectly in the agroecology and food sovereignty movements of their respective countries. Currently, they are also succeeding in making ends meet as small-scale agroecological farmers in the current capitalist food system. It was anticipated that this particular positioning could yield the tensions and contradictions between farmers' immediate economic survival (which entails working within the existing system) and their other goals and aspirations, such as regenerating their land, building communities and equitably nourishing populations, which may require transformation of the systems on which they currently depend.

3. Approaches identified by farmers through participatory processes

In this section, we summarise the different framings that farmers used to depict their food system dynamics, and the strategies identified to transform them towards agroecological food systems. As an action research project, the framings and strategies proposed evolved over the course of the project, and significant changes in framings are noted.

Key themes that emerged from the research, deliberation and development of strategies in relation to the research question included approaches to:

1. appropriately value agroecological food and farming;
2. increase access to land for agroecological farmers;
3. ensure imports and exports are fair to producers and consumers across countries;
4. support alternative distribution and retail that enables local and agroecological production.

Producers in Nicaragua and England were largely convergent in many of these areas, although divergences did occur and are noted in the discussion of each of these approaches.

3.1. Valuing agroecological food and farming: roles of consumers, citizens and the state

Agroecological farmer panels in Nicaragua and England agreed on the fact that, at present, prices received by farmers and paid by consumers do not adequately respond to the actual costs of food production or the wider social and ecological effects of farming approaches, what farmers in England referred to as 'true costs'.[2] During early workshops, farmers in England explicitly discussed the ways in which they felt that their work was undervalued, particularly in financial terms but also in other social terms, by both the public and the government. Similarly, in Nicaragua, agroecological farmers expressed that they did not receive a price that reflected the regenerative nature of their farming, but rather were dependent on insecure and volatile commodity markets that often priced their products below their costs of production.

Later in the research process, during the deliberative workshop in England, farmers embarked on a lengthy discussion of the concept of 'value' itself, who determines it, and what it does and does not include. Farmers concluded that there were certain things that simply could not be monetised, most notably land but also seeds and food. Despite this acknowledgement that not everything could or should be converted into a quantifiable (economic) value, they indicated a need for market transactions to incorporate the 'true costs' (water pollution, obesity and diabetes, loss of biodiversity and so on) and the 'true benefits' (healthy diets, lively rural communities, healthy soils, etc.) of food production. The panels investigated potentials of various approaches that might do this, including the roles of certifications, consumers and the state. We explore these further in the subsections below.

3.1.1. Roles and limitations of consumer behaviour

Farmers in both Nicaragua and England determined that there was a role of consumers as market actors in acknowledging the benefits to their own health, their ecosystems and their rural communities by virtue of choosing to consume agroecological rather than conventional products. In England, for example, farmers noted that supermarkets are susceptible to mass customer pressure. In Nicaragua, farmers themselves decided to pledge to personally take more responsibility for their purchasing choices.

However, as the project developed, the farmer panels, particularly the England panel, also saw the limitations of consumers in terms of market actors. Whilst initial framings of the problem towards the beginning of the research initiative focused on consumers' willingness to pay (e.g., 'if only consumers were more knowledgeable about what went into their food'), as the process went on, farmers acknowledged that they themselves often made purchases that did not support local and agroecological production. Farmers in England indicated that their choices were often driven by the budgets they had available to them (which were often cited to be relatively low based on their earnings), social pressures experienced by their children (e.g., for packed lunches at school where children compare lunches with one another) and availability of products from which to choose (i.e., the lack of availability of local, ecological and indigenous foods in their shops and/or at certain times of day).

Farmers in England also discussed the limitations of consumer knowledge and the complexity of the issues. One participant stated, 'tomatoes from Spain, pasture-fed beef … . It is very complicated, what's better? It is bewildering to consumers and farmers'. They concluded that it was important for consumers to have more information; however, it is not realistic to allocate all the responsibility to consumers and expect them to understand all the issues, and to have the financial and social positioning to consistently purchase 'ethical' products.

They noted that citizens are burdened with the responsibility for 'demanding' a sustainable food system, when they have limited means and tools to do so. At the same time, governments and the private sector relinquish this responsibility. What emerged in the participatory research was the need for consumers, especially those in urban areas, to reconnect with the origin of their food, and to mobilise politically as consumer groups, as *one of many* strategies to transform food systems.

3.1.2. Niche markets, price differentiation and their limitations

The majority of the panel members in England currently benefit from price differentiation in their business models. Price differentiation describes the ability to market goods to different markets at different prices, in this case a niche market for affluent buyers of organic and/or local products, a market that is unaffordable for the majority of the consumers.[3] The English panel highlighted how this 'dual system' (i.e., organic/non-organic) was flawed, as a promotion of agroecological production in these conditions would mean crowding and competition in the niche market (putting their own livelihoods at risk) rather than agroecological products being consumed more widely. They stressed that the goal should not simply be to make agroecological food available, but rather make *all* available food agroecological and accessible to all.

In Nicaragua, farmers discussed the paradox of vying for an organic export market that did indeed give a price premium, but in which products were exported and were unavailable to Nicaraguans, with a ceiling on volume and complex standard requirements that meant only a few large-scale farmers would benefit. The farmer panel concluded that the goal was to move the local and national markets towards agroecological products that are paid fairly, rather than depend on organic production for elite foreign consumers. Nicaraguan farmers' strategies included agroecological branding and the creation of agroecological market fairs or stalls. However, this is at odds with concerns about the limits of niche markets highlighted by the English panel.

3.1.3. A need for regulating against destructive agriculture

In both countries, farmers recognised that the state has a role and responsibility to penalise destructive unsustainable practices (e.g., water pollution and soil degradation) carried out by mainstream agriculture. In England, farmers also recognised the carbon emissions related to the transport of food, particularly from abroad. Therefore, farmers in both countries concluded that incentives towards sustainable production must come in parallel with tax and legal mechanisms that discourage unsustainable production. The legal mechanisms often exist (anti-monopoly law, environmental regulation); however, they need to be enacted to effectively discourage unsustainable farming practices.

In England, the concept of 'polluter pays' mechanisms was discussed. However, it was noted that this often becomes a system of 'those who can pay pollute'. Therefore, the panel concluded that there needs to be more strict regulation that can effectively discourage and prohibit unwanted practices.

Members of both farmers' panels saw value in linking with legal experts and law universities to find appropriate tax and legal mechanisms for dissuasion, and to put cases against corporations in events where existing environmental regulations were breached. In Nicaragua, despite excellent environmental legislation, regulations are rarely enforced. One of the strategies developed by the farmer panel in Nicaragua was the design of a practical guide for social movements to build legal cases where environmental legislation is not enforced.

3.1.4. Public interest and rights framings

As deliberations continued in England, panel members identified a need to go beyond market framings to consider support for agroecological food systems to be in the public interest. With a public interest approach, grounded in UK legislation, farmers believed that health, dignified jobs and ecosystem services (e.g., biodiversity, water quality and flood reduction) could be better taken into account, tipping the balance in favour of local and agroecological production over mainstream farming practices. It was suggested that reframing agroecological food systems in this way could both build alliances with complementary citizens' groups and create legal traction among public authorities, which are charged with promoting the public interest. In the case of Nicaragua, farmers called for the government to recognise their right to live in a healthy environment, free from pollution, a right enshrined in the national Constitution and environmental legislation. The panel determined that agroecological food production would contribute to the realisation of that right.

Overall, the focus in England significantly shifted from one that relied on consumer choice of products to a focus on building alliances between farmers and consumers as citizens.[4] While farmers recognised some role for consumers to contribute to regenerative food systems, the main emphasis became one of developing a focused effort to *collectively* define the values that should shape our food systems, including both producers and consumers in those discussions. Initial national-level efforts to engage citizens in food policy decisions were undertaken around the time of this project (Butterly and Fitzpatrick, 2017).

3.2. Increase access to land for agroecological farmers

Farmer panels in Nicaragua and England both highlighted lack of access to land as one of the key problems for transitioning towards agroecological production. In both countries, panellists recounted how new entrants are unable to get a foot in the door due to the high initial costs of purchasing land, and due to the lack of suitable land being available on the market (e.g., large parcel sizes in England). In both cases there has been a significant hike in land prices in recent years, with high levels of capture by wealthy elites and large agribusinesses. In Nicaragua, land grabs by both foreign and national investors are mostly for export-commodity production such as sugar cane, coffee, or cattle. In England, land distribution has been highly unequal for centuries, with over two-thirds of agricultural land owned by less than 1% of the population.[5] Farmland is becoming even more inaccessible to small-scale producers due to the expansion of nearby large farms (as a response to area-based subsidies and productivist business strategies), amenity use (e.g., horses) and speculative investment.

Succession and land transfers are also a problem, but for different reasons in England and Nicaragua. The tax regime in England encourages investment in land and discourages land to be either sold or transferred whilst the landowner is alive; this is because tax breaks exist when farmland is inherited. In Nicaragua, the concern is different: intergenerational land transfers are discouraged because of the need to keep children close in the final years of life.

Solutions proposed by farmer panels to address the inadequate access to land for agroecology in the two countries included strategies to address inheritance issues, issues with titling and transparency, planning issues and land grabs. These are discussed in turn below.

3.2.1. Address inheritance of land

The farmers' panels advocate that inheritance law should not contribute to further concentration of land but facilitate the ability of new entrants and younger generations to purchase land to start an agroecological farm. In the case of Nicaragua, the panel articulated a need for cultural change in which parents would progressively give their successors more autonomy as a form of inheritance in life, coupled with state policies that enable access to land for new entrants. In England, farmers indicated a need to change policies to disincentivise using land as a speculative investment and for avoiding inheritance taxes. Overall, the farmer panel in England indicated that it would be important to adjust taxation in order to favour small-scale and agroecological agriculture, but that the panel members themselves did not have the expertise to suggest concrete ways in which this could be done. There was also a sense that such reform would be relatively difficult in terms of attaining political will. However, the farmers agreed to hold a discussion with political parties and key experts in taxation approaches in order to communicate their perceived need for reform and explore the options available. At the time of writing, some project members, including those brought in later as 'allies', have been further involved in land taxation discussions with major political parties.

3.2.2. Improve land titling and tenures, increase transparency and introduce state land banks

In both countries, panellists identified a need to make land ownership and land exchanges more transparent. In Nicaragua, a plot can have several competing land titles, and land registries are incomplete and underfunded. In England, the registry of land ownership is not complete or publically available, so it is difficult to know who owns what. Similarly, when land is sold, the transfer is not made public, with sales often occurring behind closed doors, and land often sold to relatively large landowners. The panels recommended the introduction of transparent and accountable mechanisms that link land sellers and buyers. State support for the sale of land for agroecological farming was one of the key recommendations: with either funding to replicate the examples of land purchases for cooperative farming such as the Ecological Land Cooperative in England or 'land banks' in Nicaragua. These are models in which a body (a cooperative or the state, respectively in these cases) purchases land and either leases or sells it to producers at rates below market average. The panels demanded that a transition towards agroecological production should be a requirement in such schemes. In England, the panel also called for legal mechanisms to curb land speculation.

3.2.3. Collaborate with planning authorities to recognise and support small farms

Farmers in England identified the need for educating and informing planners about different interpretations of planning rules that could favour, rather than hinder, agroecological farming. Suggestions for this included, for example, greater recognition for farmers who are producing food and earning their living off the land at smaller scales than average and using agroecological approaches not yet adopted in conventional farming communities. Such recognition would help improve the chances for agroecological farmers to obtain planning permission to build homes and necessary infrastructure on their farms, which would increase their economic viability. Farmers also discussed changes in the implementation of planning policy which would require new urban development to allocate space for community (agroecological) farms and gardens.

At present, farms in England need to prove their business 'viability' in order to obtain planning permission; this excludes many low-income farms that earn enough for subsistence but do not turn a profit. In a summit that brought together the English and Nicaraguan farmer participants, the latter were amazed about the distinction in England between agricultural land and land on which people can live, as Nicaraguans build homesteads where they farm. England's farmers expressed a desire to enable people to live on non-commercial 'smallholdings'. However, within the existing planning and agricultural system in England, they also felt compelled to differentiate themselves, productive farmers, from so-called 'hobby farmers', i.e., those who might have a pony or 'a couple of heads of kale', but who are not significantly contributing to the food system. This has oriented the struggles of farmers in England towards making a case for their legitimacy as 'viable' food producers, rather than advocating for the right for people to live on the land.

3.2.4. Address land grabs

As mentioned above, land is being progressively concentrated in the hands of wealthy commodity producers in Nicaragua. This land is then used to produce in unsustainable ways, with high use of agrochemicals and water in the case of sugar cane or tobacco, or remotely without engaging with the communities where they purchase their land in the case of cattle. Similar dynamics of land concentration patterns were also identified by the farmer panel in England. The Nicaragua panel recommended that existing environmental law should be used to hold these companies to account, in parallel with efforts to allocate land specifically to small-scale agroecological producers and support their production.

Given the strong link between land use and farming practices, the farmers in England looked at policies that would 'punish' or make environmentally destructive farming practices more expensive, and subsidies that would actively encourage agroecological approaches. It was noted during the systems mapping workshop in England that there was a need to further democratise land use, particularly in relation to publicly owned land (e.g., county council farmland estates).

3.3. Ensure imports/exports are fair to producers and consumers across countries

Farmers in both countries indicated that it would be important for people to purchase food more directly and more locally. They indicated a role for imports and exports, but a much smaller one than currently exists, with consideration about what is imported and exported, how that's decided and why.

In Nicaragua, the impact of the Central America Free Trade Agreement (CAFTA) between Central American countries (including Nicaragua) and the United States has been highlighted by the panel as a major source of small-scale farmers' vulnerability. Farmers indicated that cheap imports are destroying local industries, and small-scale producers do not reap the benefits of selling in export markets. Intermediaries capture the export price markup (SIMAS, 2012). Further, international trade companies curtail price transmission: when farm gate prices are low, these low prices are transmitted to producers, but when consumer prices are high, trade companies pocket the difference rather than transmitting a higher price to producers (McLaren, 2015).

Farmers in Nicaragua see value in accessing export markets, but not in return for sovereignty. Nicaraguan farmers indicated that they see their local and national economy as they see their household economy; things that can be produced at home are not purchased outside, and one only purchases items that one does not produce (Ripoll, 2018). This is in contradiction of free trade ideologies such as those enshrined in CAFTA and other free trade policies that currently shape Nicaragua's food markets.

Farmers in England had a similar perspective about local versus international markets: local food systems were agreed to be favourable both for producers and consumers in the UK, while imports and exports were also seen as beneficial under the right circumstances. While there was no explicit discussion about free trade policies in the England farmer panel, it was noted that imports from countries where labour is cheap were a problem for farmers in England. Farmers felt that the country had chosen to specialise its food economy into grain and livestock production, ignoring key subsectors such as vegetable production that the UK entrusts other countries to produce and export to them. Farmers also noted that UK imports and exports should not undermine the sovereignty of farmers in the exporting or importing countries. Farmers indicated that the UK should only import goods that could not be feasibly produced in the UK.

3.4. Support retail and supply chains that benefit local and agroecological producers

In addition to import and export decisions, farmers also discussed a need to support 'alternative' retail and supply chains that would better support farmer livelihoods, increase transparency and allow for more 'local' transactions, while also ensuring that healthy, agroecologically produced food was accessible to all people, and not just middle-class or 'conscious' consumers.

'Local' was understood by the farmers as a closer and more transparent connection between consumers and producers, where products are adapted to the local ecology, and where a local economy enhances local life.

The capture of value through extensive chains of intermediaries in Nicaragua, or supermarket chains in England, means that a closer connection between consumers and producers, including those in urban areas, is often emphasised through the support of local markets and farmers markets, as well as the legal and financial support of alternative marketing arrangements such as Community Supported Agriculture (CSA), veg box schemes or Internet retailing such as the Food Assembly in England. However, as discussed earlier in this chapter, such approaches were deemed problematic in that they tend to create niche markets.

To address these challenges, the farmers identified different market and supply chain arrangements that could improve the current situation.

3.4.1. Support and spread knowledge about 'alternative' and cooperative market approaches

Farmers in England indicated that they did not have sufficient awareness of alternative food marketing approaches that could support local production and consumption. As part of the research and deliberation, Tamar Grow Local (TGL) was explored as a case study of urban-rural connections. TGL is a community interest company (CiC) with several interlocking elements covering the Tamar valley in Devon and the city of Plymouth. It addresses issues of distribution of resources, including aggregation of small-scale products, joint use of processing equipment, shortened supply chains, an explicit focus on providing local food to low-income populations through partnership with local government, and a limited premium on products in order to both maximise price for the producers and maintain affordability for consumers. Farmers in England were impressed with this model and took action to spread awareness of this approach so that it might be replicated in other parts of the country.

Farmers in Nicaragua also saw cooperativism and association as important in helping to support agroecological producers, both in making the purchase of inputs and services cheaper, and through planning production and pooling produce to sideline intermediaries and reduce the shock of fluctuating prices.

3.4.2. Promote the consumption of agroecological food in public institutions

Farmers considered public procurement as another possibility for urban sourcing from and promotion of agroecological producers. Hospitals, the military, school feeding programmes, etc. can give priority to agroecological produce when allocating public contracts (Swensson, 2015; Soldi, 2018). It is acknowledged that there would be periods of transition, but in the long run it would ensure a steady demand for agroecological produce and would allow for agroecological value chains to establish themselves. As in the case of the Fame Zero programme in Brazil, these new public procurement systems can be used to deliver healthy food as part of social protection programmes for the vulnerable. In Nicaragua, farmers indicated that the sourcing of grain by the public company ENABAS[6] could potentially be switched to sourcing agroecological produce from Nicaraguan farmers. ENABAS guarantees floor prices to small-scale producers and sells at subsidised prices to poor consumers.

4. Analysis and discussion

In the previous section, we laid out the analysis of the farmer panels in both Nicaragua and England, and the strategies that they have proposed to transform food systems into agroecological or regenerative food systems. The implications of this are discussed below.

4.1. Concomitance of different approaches

Holt-Gimenez and Shattuck (2011) classify strategies to change food systems as neoliberal, reformist, progressive or radical, and indicate that the former two, neoliberal and reformist, are unlikely to sufficiently address the nutritional, ecological and social justice deficits of existing food systems. They argue for greater alliances between progressive and radical strategies, with the latter essential as they explicitly aim to undermine the capitalist relationships that underpin many of the problems of existing food systems. Within this framework, the strategies proposed and implemented by Nicaraguan and English farmers in this project included a significant emphasis on progressive and radical strategies as well as some reformist approaches. This is illustrated in Table 8.1.

In their discussions, farmers actively recognised the tensions that exist between working within and transforming food systems, with explicit concerns about co-option, and being trapped into a niche market. However, they also acknowledged concerns about feasibility, political will and insufficient capabilities, and the expertise needed to enact transformative changes to food systems. Therefore, farmers identified a need for a combination of reformist, progressive

TABLE 8.1 Political orientation of farmer panel analyses and strategies

Politics	*Farmer panel analysis and strategies*
Neoliberal	• None
Reformist	• Changing subsidies to favour agroecological farms in England • Certification of agroecological products in Nicaragua • Working with planning officers to improve the ability for agroecological farmers to live on farmland • Discouraging destructive agricultural practices
Progressive	• Promotion of cooperatives to enable producers to have a greater share of food value and better access to equipment • Policies to increase access to land for new entrants, young farmers, and small-scale producers in England via land cooperatives, and in Nicaragua via market-based land redistribution (land banks) • Improving transparency in food and land markets • Promotion of market approaches that favour agroecology (e.g., quotas for agroecological sourcing in public procurement)
Radical	• Demanding a departure from perceiving food and land as commodities • Promotion of public interest and rights framings to go beyond production and consumption to wider considerations of purpose and approach • Promotion of market approaches that prioritise collaboration over competition through, e.g., the Tamar Grow Local model that coordinates producers and links to low-income consumers • Democratising farmland through collective purchasing and ownership by local authorities • Addressing land grabs • Reducing imports to those that are necessary and reorienting production towards food needs, with a view to protect the needs of people in other countries

Source: Authors.

and radical approaches in the short term; none of them were sufficient on their own. However, the coexistence of radical and progressive strategies had one key requirement – that they are combined in ways that powerful actors do not co-opt them and in ways that they do not undermine their long-term goals.

It is noteworthy that while some farmer strategies did conform to capitalism, many of the strategies that farmers proposed, and are engaged in implementing, are actively countering various aspects of capitalism (see Chapter 10 by Tilzey). Farmer strategies included:

- reuniting workers with their means of production (through increasing access to land);
- undermining the dominance of private property (albeit through minor strategies of public and collective land ownership);
- reducing competition amongst producers (through cooperatives and non-competitive public tendering);
- refusing to separate the economic from the political (e.g., in reframing food production in non-economic ways);
- countering commodity fetishism (e.g., through UK farmer rejection of dual markets as a solution, and through rejecting land and food as commodities in general);
- challenging the predominance of exchange value over use value (e.g., in 'public interest' and rights framings).

4.2. Shifts in discourse throughout the participatory process

Farmer participation in action research, and particularly in a study that reaches across the whole food system, changed their focus of concern from issues around individual agency to problems that are more structural and further removed from the farm, and at times taking on issues that no longer affected the participants directly (e.g., policies related to access to land advocated for by farmers who had already obtained their planning permission). Overall, there was a shift throughout the process from a focus on sub-hegemonic or 'reformist' (Holt-Gimenez and Shattuck, 2011) framings (e.g., 'let's take away the subsidies of the giants and see how they fare'; and 'if consumers knew more they would change') to those that are congruent with food sovereignty or 'progressive' and 'radical' approaches (e.g., efforts to curb land grabbing, promotion of non-competitive markets, cooperative land schemes). The reflexivity, deliberation and an iterative design of this project that led to shifts from reformist to progressive and radical strategies are relevant for future participatory projects looking to transform food systems.

4.3. Similarities and differences in strategies in Nicaragua and England

Despite a potential anticipation that farmer-activists in the Global South would be more radical (Chapter 10 by Tilzey), the strategies amongst farmers in

Nicaragua and England as part of this project were relatively equal in terms of their distribution between reformist, progressive and radical politics.

Imports and exports are significant ways in which producers and consumers in the Global North and Global South are affected by one another. Farmers in both Nicaragua and England called for the management of imports and exports according to the goals of food sovereignty rather than capitalist market logic. This could, if it were realised, reduce the 'imperial mode of living' in which social and ecological surplus is expropriated from the Global South to the Global North (Brand and Wissen, 2012). Rather than causing problems in the Global North, farmers in England articulated ways in which this change could improve livelihoods, health and ecosystems. However, this might have been different if the groups had been consumers, rather than producers, and/or if perspectives were not challenged through the participatory process.

There were, however, significant points of divergence between the two panels. Farmers in England advocated for the recognition of agroecological farmers as 'viable' and 'productive' businesses. The Nicaraguans questioned this strategy and the lack of advocacy for all people to live on the land, regardless of whether they were 'peasants' or 'viable businesses'. Arguably, the focus of the England panel stems more from pragmatism than from political stance. Working within the existing planning policy framework was seen as preferable to attempting to challenge one of its main underpinnings: the separation of agricultural land from housing land. It also stemmed from concern amongst panellists that eliminating this distinction would further increase the price of land and decrease its availability. In terms of politics, though, this strategy runs a risk of agroecology being co-opted into the entrepreneurialism that results in ecological and social trade-offs that the panel was so strongly against.

A second point of divergence related to certifications. The panel in England firmly concluded that product differentiation was a limited and flawed strategy and that, instead, efforts needed to be made to convert all food production to agroecological methods. In Nicaragua, however, farmers advocated for the differentiation of their agroecological produce from that which was conventionally produced. This was seen as a pathway for specifically supporting agroecological producers, possibly through ensuring that their products are preferentially purchased by consumers. It was also seen as important for engaging consumers in their purchasing decisions.

4.4. Urban obligations and the need for democratisation

A transformation of food systems requires not only changes in agriculture, but also changes in the ways urban markets and public institutions choose to source food, and in turn, what responsibilities urban consumers are perceived to have. Retail and manufacturing rely on just-in-time delivery and demand seasonal products all year round. They therefore source from complex and often global value chains rather than the local markets and short food circuits that are more amenable to agroecological production. Public procurement of food is often allocated in terms of price, rather than for demonstrating sustainability and local sourcing. Similarly, planning policy focuses either on urbanisation or on saving 'green' areas, instead of also catering for sustainable forms of agriculture and facilitating a closer contact between people and where their food is produced. The application of agroecological food systems in rural-urban contexts emphasises the necessity for diversification, zoning rural-urban landscapes, planning for seasonality in a food systems context, and producing at scale (Vaarst et al., 2018). City dwellers perceive themselves to be consumers in linear value chains, rather than as interdependent participants (food citizens) with rights and obligations in how food systems should be governed (Food Ethics Council, 2017). Indeed, farmers in the panels pointed towards the detachment of consumers from the land and food production as a major challenge. They saw a need to engage people not as consumers but as food citizens who needed to be on board in order for land distribution and localised food to deliver more agroecological food systems. Further, they would need to see the countryside in a different way: farmers in England noted how nimbyism[7] exists on a national scale; many people prefer to import food from abroad and have the countryside in their own country conserved for recreation and aesthetics. Reconnecting people to food and land was seen as essential for ensuring that efforts to democratise food result in shifts towards agroecology.

5. Conclusions

Our participatory research demonstrated the potential for agroecological farmer-activists in the Global North and Global South to apply a diversity of approaches to transform their food systems. Whilst some strategies embraced radical approaches, other strategies sought to find pragmatic solutions within the current system. Overall, farmer panels valued a combination of reformist, progressive and radical strategies. Ultimately they determined that none of the strategies were sufficient on their own: a combination was necessary. An articulation of common values to underpin food systems was seen as essential, as was working to redistribute land (albeit through the existing system) and creating alternative markets that were based on collaboration rather than competition, and which fulfilled the public interest (in England) and human rights (in Nicaragua). Alliances with urban consumer-citizens and public institutions (city councils, urban-rural planning, public procurement, etc.) were seen as a necessary complement to land access and local food circuits for a transition into agroecological food systems. Pragmatism, both about current political openings and the panel members' capabilities, guided which strategies were taken forward, although the majority of strategies developed, including progressive and radical, are being carried forward.

The project confirms that farmer-activists, both in the Global North and the Global South, and even those currently engaged in and benefiting from markets, are capable of advocating for and developing strategies that could lead to radical transformations of our food systems. It is particularly worth noting that in many situations, strategies proposed by farmers went beyond their individual interests. This can partly but not entirely be attributed to the reflexive, deliberative approach employed. It also indicates that reflexive, iterative and deliberative approaches can help in further clarifying and teasing out contradictions, which in this case led to a deeper challenging of capitalist underpinnings of food systems.

Notes

1 Their classification includes neoliberal, reformist, progressive, and radical approaches.
2 In England, some of the farmer panellists were aware of an initiative called 'True Cost Accounting', led by the Sustainable Food Trust, and the phrasing of 'true costs' could have come from this initiative.
3 For example, organics now comprise a larger proportion than ever of food markets; however, organic food is still the minority (only 1.5% in the UK, based on financial value) (Soil Association, 2018).
4 The farmer group in England often used the term 'citizens' but in recognition that many people live and eat in England without necessarily being a citizen, the panel also used the term 'people'.
5 Based on Defra 2016 data on agriculture and forestry holdings, and supported by figures in Cahill (2011). Thank you to Robin Grey, who has worked on these calculations.
6 Empresa Nicaragüense de Alimentos Básicos.
7 'Nimbyism' refers to 'Not in My Back Yard', objecting to something unpleasant or hazardous in one's own neighbourhood but not elsewhere.

References

Altieri, M. (1987) *Agroecology: The Science of Sustainable Agriculture*. Westview Press, Boulder.

Altieri, M. (2009) 'Agroecology, small farms and food sovereignty', *Monthly Review*, vol 61, no 3, pp102–113.

Bernstein, H. (2014) 'Food sovereignty via the 'peasant way': A sceptical view,' *The Journal of Peasant Studies*, vol 41, no 6, pp1031–1063.

Brand, U. and Wissen, M. (2012) 'Global environmental politics and the imperial mode of living: Articulations of state-capital relations in the multiple crisis', *Globalizations*, vol 9, no 4, pp547–560.

Burns, D. (2012) 'Participatory systemic inquiry', *IDS Bulletin Special Issue* 'Action Research for Development and Social Change', vol 43, no 3, pp88–100.

Burns, D. and Worsley, S. (2015) *Navigating Complexity in International Development*. Practical Action Publishing, Rugby, UK.

Butterly, D. and Fitzpatrick, I. (2017) *A People's Food Policy*. www.peoplesfoodpolicy.org

Cahill, K. (2011) *Who Owns Britain and Ireland*. Cannongate Books Ltd, Edinburgh.

Carpini, M. X. D., Cook, F. L. and Jacobs, L. R. (2004) 'Public deliberation, discursive participation, and citizen engagement: A review of the empirical literature', *Annual Review of Political Science*, vol 7, pp315–344.

Chappell, M. J. and La Valle, L. A. (2009) 'Food security and biodiversity: Can we have both? An agroecological analysis', *Agriculture and Human Values*, vol 28, no 1, pp3–26.

Desmarais, A. A. (2007) *La Vía Campesina: Globalization and the Power of Peasants.* Fernwood, Halifax.

Edelman, M. (2014) Food sovereignty: Forgotten genealogies and future regulatory challenges, *The Journal of Peasant Studies*, vol 41, no. 6, pp959–978.

Food Ethics Council (2017) *Food Citizenship. How thinking of ourselves differently can change the future of our food system.* Findings of the Future of Food: a collaborative innovation project. Food Ethics Council, London. Available at https://foodcitizenship.info/.

Gliessman, S. R. (2007) *Agroecology: The Ecology of Sustainable Food Systems.* CRC Press, Boca Raton.

Gonzalez de Molina, M. (2013) 'Agroecology and politics. How to get sustainability? about the necessity for a political agroecology,' *Agroecology and Sustainable Food Systems*, vol 37, no 1, pp45–59.

Holt-Giménez, E. and Shattuck, A. (2011) 'Food crises, food regimes and food movements: rumblings of reform or tides of transformation?', *The Journal of Peasant Studies*, vol 38, no 1, pp109–144.

La Via Campesina (2007) *Nyéléni 2007: Forum for Food Sovereignty.* Sélingué, Mali.

Levidow, L. (2015) 'European transitions towards a corporate-environmental food regime: Agroecological incorporation or contestation?', *Journal of Rural Studies*, vol 40, pp76–89.

McLaren, A. (2015) 'Asymmetry in price transmission in agricultural markets', *Review of Development Economics*, vol 19, no 2, pp415–433.

Meadows, D. (1997) 'Places to intervene in a system: Strategic levers for managing change in human systems', *Whole Earth Review*, vol 91, no 1, pp78–84.

Meadows, D. (2008) 'Leverage points: Places to intervene in a system', in D. Meadows (ed), *Thinking in Systems: A Primer.* Chelsea Green Publishing, Sustainability Institute, Hartland.

Mendez, V. E., Bacon, C. M. and Cohen, R. (2013) 'Agroecology as a transdisciplinary, participatory and action-oriented approach', *Agroecology and Sustainable Food Systems*, vol 37, no 1, pp3–18.

Patel, R. (2011) 'What does food sovereignty look like', in H. Wittman, A. A. Desmarais and N. Wiebe (eds), *Food Sovereignty: Reconnecting Food, Nature and Community.* Pambazuka Press, Oxford.

Pimbert, M. P., Thompson, J. and Vorley, W. T. (2002) 'Global restructuring, agri-food systems and livelihoods'. *Gatekeeper Series 100.* International Institute for Environment and Development, London.

Pretty, J. N., Noble, A. D., Bossio, D., Dixon, J., Hine, R. E., Penning de Vries, F. W. and Morison, J. I. (2006) 'Resource-conserving agriculture increases yields in developing countries', *Environmental Science and Technology*, vol 40, no 4, pp1114–1119.

Reason, P. and Bradbury, H. (2008) *The SAGE Handbook of Action Research: Participative Inquiry and Practice.* Sage Publications, London.

Ripoll, S. (2018) *As Good as it Gets? The New Sandinismo and the Co-Option of Emancipatory Rural Politics in Nicaragua*, ERPI 2018 International Conference Authoritarian Populism and the Rural World, Conference Paper No.46.

Senge, P. M. (1990) *The Fifth Discipline.* Currency Doubleday, New York.

Sevilla Guzmán, E. and Woodgate, G. (2013) 'Agroecology: Foundations in agrarian social thought and sociological theory', *Agroecology and Sustainable Food Systems*, vol 37, no 1, pp32–44.

SIMAS (2012) *Mercados campesinos. Más que un lugar para comprar y vender.* PCaC-UNAG, MS-Action Aid and SIMAS, Managua.

Soil Association (2018) 'Organic market 2019', available from https://www.soilassociation.org/media/18224/omr-report-2019-interactive.pdf (Accessed August 2019).

Soldi, R. (2018) *Sustainable Public Procurement of Food*. European Committee of the Regions, available at https://cor.europa.eu/en/engage/studies/Documents/sustainable-public-procurement-food.pdf#search=Sustainable%20public%20procurement%20of%20food.

Swensson, L. F. J. (2015) *Institutional Procurement of Food from Smallholder Farmers: The Case of Brazil*. Food and Agriculture Organization of the United Nations, Rome.

Tilzey, M. (2017) 'Reintegrating economy, society, and environment for cooperative futures: Polanyi, Marx, and food sovereignty', *Journal of Rural Studies*, vol 53, pp317–334. doi: 10.1016/j.jrurstud.2016.12.004.

Vaarst, M., Escudero, A. G., Chappell, M. J., Brinkley, C., Nijbroek, R., Arraes N. A., Andreasen, L., Gattinger, A., De Almeida, G. F., Bossio, D. and Halberg, N. (2018) 'Exploring the concept of agroecological food systems in a city-region context', *Agroecology and Sustainable Food Systems*, vol 42, no 6, pp686–711. doi: 10.1080/21683565.2017.1365321.

9

CONJUGATING SOCIAL AND SOLIDARITY ECONOMIES IN CHIAPAS, MEXICO

Redesigning food systems for economic, social, and ecological virtuous circles

Emilio Travieso

1. Introduction

The dominant food system, based on an industrial model of production and a market model of distribution, has failed to feed the world's growing population, despite producing more than enough food (FAO et al., 2015; De Schutter, 2014). Meanwhile, it has created new social, environmental, and health-related problems (McIntyre et al., 2009; van der Ploeg, 2010; Hoffman, 2013; De Schutter, 2014). As a result, more and more policy-makers are realising that "business as usual is no longer an option" (McIntyre et al., 2009). They are lending greater support to innovative food system designs, such as those that prioritise the role of small farmers and agroecological modes of production (cf. McIntyre et al., 2009; De Schutter, 2010; Altieri and Toledo, 2011). At its core, the challenge of building more sustainable, just, and healthy food systems requires a paradigm shift in our economic model.

Among the most promising experiments are those that combine agroecology, food sovereignty, and social and solidarity economy. Agroecology, which in its contemporary form was first developed in Mexico (Hecht, 1995), is an approach that applies ecological principles to agriculture. It has grown in sophistication since the 1970s, incorporating scientific and traditional knowledge together with concern for sociocultural dimensions; its scope as a multidisciplinary, action-oriented approach has expanded from the field level to its present concern with entire food systems (Hecht, 1995; Francis et al., 2003; Méndez et al., 2013). Its proponents argue that sustainable food systems must not be modelled on modern industry, but rather on nature; they must be cyclical, not linear (Altieri, 1995; van der Ploeg, 2010: 99-100; Jones et al., 2011).

Agroecologists also promote local food systems. This would reduce dependence on the fossil fuels used to transport inputs and outputs (De Schutter, 2010:

13), while guaranteeing that food is fresh, with less need for additives that increase shelf life but can be unhealthy. Further, local food systems are more likely to be embedded in relationships of mutuality, which, in turn, will have an effect on distribution and access.

These factors are related to the food sovereignty paradigm, which posits the right of peoples to democratically determine the policies guiding the production, trade, and degree of self-sufficiency in their food systems, rather than being at the mercy of market mechanisms or other external forces that would determine this for them. Because these innovations call powerful regimes into question, they are manifestly political (González de Molina, 2013; Holt-Giménez and Altieri, 2013).

Social and solidarity economy (SSE) is another vibrant field of social innovation. Initiatives in SSE redefine the terms of economic exchange. In general, solidarity economy emphasises reciprocity and mutuality over profit, while social economy tries to make business useful to social objectives. The terms, and the related but distinct movements they represent, have been joined under the SSE umbrella partly as a political strategy to increase the visibility of both (Dacheux, 2013). Like agroecology and food sovereignty, SSE has grown in both complexity and scope, notably around the World Social Forum that first met in Porto Alegre, Brazil, in 2001 (Hart et al., 2010).

Many initiatives combine some or all of the above innovations. However, our understanding of their deeper complementarities has not been sufficiently developed (cf. Spratt, 2015). This chapter aims to contribute to that understanding through an empirical case study, based on seven months of ethnographic fieldwork and ongoing participatory action research.

2. The Misión de Bachajón in Chiapas, Mexico

The Misión de Bachajón (MB) constitutes the local Catholic Church in a 3,000 km^2 area of the northern Selva (Jungle) region of Chiapas, which is Mexico's southernmost state. It is an especially innovative organisation from an especially marginalised context.

Chiapas has a population of roughly 4.8 million people (INEGI, 2010), of which 74.7% was living in multidimensional poverty in 2012, the highest rate of any state in Mexico (CONEVAL, 2012). Indigenous people in Chiapas have the lowest Human Development Index value, rated at 0.61, among any of Mexico's 31 states or federal districts (MRGI, 2011).

The largest indigenous ethnic group are the Tseltal, whose language has over 461,000 speakers, followed by the Tsotsil, another Mayan group with some 417,000 speakers; in addition, there are 10 other federally recognised groups, most of which are also Mayan. Indigenous peoples in Chiapas have been the object of urbanisation programmes since the 16th century (cf. Vogt, 1969: 22–23) and as recently as 2007 (Wilson, 2011). Despite this, 51% of the total population of Chiapas is rural, compared with 22% nationally (INEGI, 2010). The proportion of rural dwellers is much higher among indigenous people.

The MB was founded in 1958 as a Jesuit mission to the Tseltal people. Its territory comprises the municipalities of Chilón and Sitalá, and some rural parts of Yajalón, Ocosingo, Simojovel, Pantelhó, Salto de Agua, and Tumbalá. The population of 300,000 is about 98% Tseltal; roughly 80% are Catholic (Zatyrka, 2003). It is spread out in more than 650 rural villages, as well as a few small towns where non-indigenous people also live.

Starting in the 1960s, the MB embraced the spirit of the Second Vatican Council, which had two major implications. One was the recognition of indigenous culture, with a corresponding process of decolonial 'inculturation'. The MB adapted itself to indigenous modes of expression and organisation. It made efforts to revitalise the Tseltal language and worldview. It adopted many traditional rituals, including those that are attuned to nature with regard to the agricultural cycle.

Through this process, the MB has grown into an 'autochthonous church', organised according to the indigenous *cargo* system of self-government (Zatyrka, 2003; Crispín Bernardo and Ruiz Muñoz, 2010). In the Mesoamerican *cargo* system, the community entrusts certain people with a specific service to the community. Young adults start with simple *cargos*, and insofar as they are recognised for honesty, generosity, and capability, they may be asked to take on increasing responsibility and accorded its corresponding authority. Those who have served the community well throughout their lives eventually join a council of elders, who are the community's highest authority. The MB 'recovered' this indigenous system and made it the basis of its own ecclesial structure. Today there are roughly 20,000 *cargos* (i.e. perhaps 20% of all adults in the MB's territory) who are engaged in over 25 types of service.

Simultaneously, the MB adopted a 'preferential option for the poor'. As part of the liberation theology movement, the MB sees the struggle for social justice as an integral dimension of its religious mission. It has played a key role in organising grassroots movements for land reform, human rights, and indigenous autonomy in the region over the last several decades.

In the context of the Zapatista uprising in 1994, the MB and its allies achieved an unprecedented agrarian reform (Bobrow-Strain, 2007). At that critical juncture, the MB recognised the opportunity, and the need, to design an alternative economic system that would help consolidate this watershed achievement. This would necessarily entail 'rebooting' the food system (cf. Borras et al., 2015). Indeed, the struggle for *food* sovereignty is indissociable from a wider effort to build *full* sovereignty for the Tseltal people, encompassing cultural, political, and economic dimensions, among others. To make this project viable, the MB has designed a sophisticated post-capitalist model in which two distinct economic spheres are first separated and then articulated.

3. Restoring economy's tension

When we think of the economy, we tend to think in terms of money.[1] However, if we look at the economy as everything that sustains and mediates the reproduction of life, we get a much wider picture. Economic anthropologist Stephen Gudeman (2008) has developed a model that accounts for economy in this sense.[2]

For Gudeman, every human economy consists of two fundamental spheres: market and community. The market is that sphere in which we interact as individual rational choosers exchanging goods and services. The community, in contrast, is the sphere of relationships where we share what we have, with a logic of solidarity. The community is founded on a common 'base', made up of those material and symbolic things that are *not for sale*.

The base is withheld from commodification because it is the guarantee of identity and survival of the community in the very long run. As such, the base has stewards, but no owners. Also, the base is composed of heterogeneous elements, whose value is incommensurable.

To be sure, the spheres of market and community are not ontologically distinct; things can pass from one to the other, the lines can be blurry, and spheres inevitably overlap (Gregory, 1997; Gudeman, 2016: 13). However, the existence of the two spheres is a social fact. Gudeman's observation, like that of other ethnographers (cf. Carsten, 1989; Little, 2004), is that people tend to organise their economic lives this way, and even go to great lengths to keep the spheres separate.

People are their 'disjoint selves' when they are in the market, and their 'conjoint selves' when they are in community. In other words, we think and act differently according to the spheres in which we see ourselves acting. Parry and Bloch (1989), who propose a similar model, add that market interactions tend to be short term, while community relationships tend to be long term.

The two spheres are connected, because they need each other. The market cannot function if there is no trust; this trust is built in the community. At the same time, the community goes to market to obtain what it needs for maintaining or renewing its base.

On the other hand, the respective logics of the two spheres (long-term sharing on the one hand, and short-term self-interest on the other) are opposed. The need to creatively negotiate this tension is the motor that ultimately drives all economies. Within this dynamism, the market tends to grow at the expense of the community and the base, in two main ways.

Firstly, it 'cascades' into the community sphere, as the frequent experience of market trade puts us in the habit of short-term, self-interested calculation; people then gradually begin to apply this to aspects of their lives that they previously treated as a distinct realm of value, governed by the logic of sharing or stewardship. Encroaching commensuration thus redraws the border between the two spheres to the advantage of the market; it shrinks the space allotted for relationships of mutuality. When the 'base' is thus overtaken by market logic, it becomes a mere commodity, at risk of alienation.

Secondly, the market sphere, by virtue of its rent-seeking logic, tends to extract value from the sphere of community-and-base (for example, in the form of unpaid family labour or productive biological processes) without nourishing it in return. The market's habits of colonisation are unsustainable because, as we have seen, a strong community and base is the condition for the possibility of a strong market. When it invades and exploits the foundation on which it rests, the market undermines the whole economy.

This is precisely the crisis we are facing; capitalism has degraded both the social cohesion and the ecological conditions that allow us to live together in our common home. The question, then, is how to regain economic balance, before it is too late. Concretely, this means restoring the creative tension between the market, with its logic of maximising short-term benefit, and the community, where persons (and the collective) are ends in themselves; at stake is the base that sustains our life.

4. Alternatives

If the unbridled market is like a wild animal, a lion, for example, that invades a village and eats the people, then the people have three options to defend their community (Travieso, 2016). The most obvious one would be to kill the lion. This has been the strategy of communist experiments, which have tried to build economies with no market. History suggests that this option is not very viable.

The second possibility is to build a fence. That is, to demarcate a protected area, where the market is not allowed to enter. This is the strategy of solidarity economy, which tries to generate a space for economic life that is not mediated by money (except perhaps a strictly local currency). A solidarity economy emphasises reciprocal practices, such as barter, and generous ones, such as voluntary service. The risk with this strategy is that when people build a fence to keep the lion out, they also lock themselves in, isolating themselves. This is why these initiatives tend to be small and fragile.

The third possibility is to tame the lion, and even domesticate it, using its strength for the ends of the community: in other words, to place the market at the service of the people, rather than the service of capital accumulation. This is the logic of social economy, whose typical expression is the 'social business'. These function with the efficiency of a capitalist enterprise, but their principal mission is to create a social good. The risk with this strategy is that it requires getting very close to the lion, and sometimes lions end up eating the people who would tame them. In other words, social economy by itself can become a Trojan horse that allows the market to sneak even further into the sphere of community, displacing other bodies such as civil society and the state (cf. McMurtry, 2013).

Gudeman's model, and the allegory of the lion, allow us to understand what is at stake in the repertoire of options available to those communities, such as the Tseltal, who aim to restore themselves to a position of strength vis-à-vis the globalised capitalist market. They will also allow us to appreciate the genius in the way that the MB has approached this challenge. To restore the health of its region's economy, and with it the possibility of a shared, sustainable life, the MB has creatively articulated two complementary strategies.

5. Protecting the base

On the one hand, the MB protects its 'base', everything that is essential for the community's reproduction, by withdrawing it from the sphere of market commodification. One integral part of the base is land. Therefore, after helping to achieve the massive agrarian reform of the 1990s, the MB promptly ensured that the regained lands would have the legal status of *ejidos*.

The *ejido* is a community-based form of land ownership, dating to the agrarian reforms of the Mexican Revolution, and especially the Cárdenas presidency of the 1930s. In the *ejido* system, land is "viewed as having a 'social function' as opposed to merely being seen as an economic factor of production" (Barnes, 2009: 394). Each family is allotted a few hectares; however, the land is owned collectively, so it can be redistributed only within the bounds of the community.

The model has been under threat since 1992, when President Salinas de Gortari introduced neoliberal policies that included changing the Constitution to allow the private sale of *ejido* land, and halting any further governmental redistribution of land. By drawing up *ejido* statutes that prohibit privatisation, and mobilising support for elected leaders who will defend this principle, the MB resists the threat of neoliberal policies aimed at turning the land into a currency of exchange. Similarly, the MB defends its territory's natural resources and biodiversity by helping to coordinate a state-wide indigenous movement that rejects extractivist and mega-infrastructure projects.

The non-material aspects of the base, such as the Tseltal language and culture, are just as important. The MB takes pains to revitalise these, together with the *cargo* system mentioned above, in which a large proportion of the adults in any given village are entrusted with a specific responsibility, ranging from religious ministry to traditional healthcare to conflict mediation. This form of work, which provides essential services and reproduces social cohesion, is paid only with social recognition, never with money.

Food is at the nexus of all of these elements. Tseltal culture is largely built around the practice of *milpa*, a Mesoamerican polyculture of maize, beans, and squash combined with a plethora of other beneficial plant, animal, and fungal species that coexist with these crops. Effectively, all Tseltal peasant families use the greater part of their land to make *milpa*, and they produce nearly everything they eat.

This degree of food sovereignty goes beyond ensuring nutrition; it entails the reproduction of knowledge, practices, vocabulary, and an entire worldview that all hinge on an intimate relationship with the local environment; it also allows for the expression of identity through special recipes whose ingredients are commercially unavailable. Food also mediates human relationships of family, hospitality, solidarity, and celebration. Indeed, the deeper political implications of food sovereignty have not been lost on indigenous scholars elsewhere (cf. Daigle, 2019).

Because of the centrality of food sovereignty to wider Tseltal sovereignty, the MB has established the role of 'caretakers of Mother Earth' (*kanal lum k'inal*). Those with this *cargo* teach their communities to expand and deepen their agroecological practices. Moreover, they emphasise the importance of using naturally 'open-source' seeds as a way of resisting the commodification of their food.

Meanwhile, Tseltal peasant livelihoods also depend on cash, and the principal source of income for most families is the annual coffee harvest. Coffee is not in any way essential to the symbolic or material 'base' of Tseltal society. It was introduced only a few generations ago, and – although it is consumed as a staple – it has little nutritional value. Therefore, it does not need to be fenced off from the market. Whereas the MB's food sovereignty strategy revolves around *milpa*, coffee (and to a lesser degree, honey) is the food product through which the MB implements a separate strategy of sovereign engagement with the outside market.

6. Taming the market

Smallholders in Chiapas are the force behind Mexico's ranking among the top coffee-producing countries worldwide (cf. Sánchez Juárez, 2015). However, the municipalities where the MB is located, which are among the most productive in the state, experience extremely high rates of multidimensional poverty relative to the rest of the country. Indigenous households in particular are disproportionately represented among the producers *and* among the poor.

The MB interprets this paradox through the lens of power. Coffee is the tropical commodity par excellence, not only because it is the most traded, but also because it exemplifies unequal relationships in global value chains. Small producers in Chiapas typically sell parchment coffee to traders, as a commodity to be roasted and brewed by third parties, mainly a small group of multinational firms, who keep the lion's share of the final price (cf. Daviron and Ponte, 2005; Irezabal Vilaclara, 2016). Financial speculation in the commodities market leaves producers vulnerable to price volatility, which is driven by factors outside their control. The fact that they are typically paid by weight takes away the opportunity to optimise value by investing in quality.

In this context, the MB eschews popular development strategies that rely solely on the conscience of consumers (such as 'fair trade' schemes), setting out instead to overcome exploitation at a structural level and restore right relationships by engaging trade from a position of strength (Irezabal Vilaclara, 2016). 'Taming' the market in this sense entails a manifold strategy of economic upgrading, focused primarily on forward vertical integration (i.e. controlling the entire value chain from the bottom-up). The MB does this through its group of cooperatives and social businesses, Yomol A'tel (YA').[3]

YA' has built its own coffee roasting plant in Chiapas, and a chain of its own brand of coffee shops located in various Mexican cities. Controlling the entire coffee value chain in this way not only creates higher returns, but also circumvents the price volatility of the commodities market. YA' is able to pay producers 30% more, on average, than the market, and to offer them a price incentive for quality. Further, YA' develops linkages to other value chains, such as honey and honey-based soaps.

Implementing this successful business strategy is necessary; however, it is not sufficient to achieve the MB's goals. The whole point is to improve conditions for the people at the bottom, and economic upgrading does not necessarily do this on its own (Barrientos et al., 2011). Therefore, the MB must not only 'tame' the market, but also 'domesticate' it.

7. Domesticating the market

Yomol A'tel's business model places the market's force at the service of the community and its base, by design. Producers earn better and more stable prices for their coffee. Since they are organised as a cooperative, they distribute income equitably and make decisions together. Profits are reinvested in the form of social property through the cooperative's microfinance institution; this institution makes loans for diversification of production, as well as for personal needs. The roasting plant, microfinance institution, and other related activities create new technical and administrative jobs for youths from producer families. As a matter of MB and YA' policy, women have equal access to these high-value and non-traditional jobs. Cross-cutting educational components, ranging from on-site training to sponsorship of university-level studies, ensure that all members grow in capabilities.

The YA' design contributes to social cohesion in several ways. Improving economic conditions for individuals makes it viable for them to offer unpaid community service through the *cargo* system. Reducing inequality and sharing power democratically helps strengthen the social fabric.

Moreover, YA' makes a structural contribution of crucial importance to the viability of Tseltal society. As the Tseltal population grows, so does pressure on the land; the only way to enable demographic growth without it resulting in poverty and emigration is through economic activities that are subject to increasing returns on investment (Reinert, 2003). This is precisely what YA' does by creating new industrial and advanced-service jobs within Tseltal territory (Travieso et al., forthcoming).

By definition, increased population density and diversified non-agricultural jobs will make northern Chiapas less rural. However, YA' avoids industrial activities that would displace or destroy traditional smallholder livelihoods. Instead, it opts for an industrial sector that depends entirely on agricultural production, with a quality upgrading strategy (described below) that incentivises agroecological intensification rather than agro-industrial expansion. Furthermore, its scalability model is based on the decentralised replication of small processing plants throughout the territory, and it preferentially hires the children of local producers for its value-adding jobs.

In this way, YA' fosters the emergence of small-town urbanism with strong rural-urban linkages rooted in multi-layered relationships. As an alternative to emigration to far-away cities, or to a more disfiguring transformation of the territory, the YA' model for economic development is respectful of Tseltal culture and society. It is also ecologically sustainable.

In contrast to exploitative "salvage capitalism" (Tsing, 2015), the YA' design establishes a relationship of mutuality with the ecosystems from which it receives services (cf. Comberti et al., 2015). Its coffee is certified organic and grown in bird-friendly shade forests. Beyond that, YA' employs a full-time agroecological engineer who, in a complementary role to the caretakers of Mother Earth, advises producers on how to integrate their coffee production with their *milpa*.

8. Virtuous circles

As we have seen, the MB's economic design keeps the spheres of market and community/base separate. The community 'protects' and restores its base (land, food, culture, etc.) with a solidarity economy approach. In contrast, it engages the market through a social economy approach to 'taming' and 'domesticating' the coffee value chain. In other words, the MB's model is fundamentally a 'middle peasant' one, in which families grow most of what they consume as well as complementary crops for exchange, but with certain 'twists' whose implications are worth exploring further.

The viability of a middle peasant model hinges on several ecological, social, and market-related factors (Li, 2015). Having achieved major structural change already, the MB understands that all of those factors are contingent. Through collective action, it is possible to create the conditions that will enable its model to survive.

The MB aims to achieve this through interconnected strategies that articulate the two spheres of its economy, creating a circuit between them that allows for virtuous circles between the ecological, social, and market-related dimensions. We have seen, for example, how Yomol A'tel's social economy strategies translate value captured in the market back to the community and its base. Likewise, the strong conjoint identities of the *cargo* system (rooted in the base) are what have enabled YA' to undertake its ambitious upgrading strategy in the market.

If today the cooperative provides a better price to producers (in the market sphere) partly in order to enable their volunteer work as *cargos* (in the community sphere), the mechanism for obtaining this higher price (forward vertical integration, etc.) is itself the result of many years of investment in which many people living in multidimensional poverty had to work even harder than usual (because of the quality controls entailed by the upgrading strategy), without any immediate gain (since higher incomes were initially offset by a 'tax' to pay back the cooperative's debts, until it finally began turning a profit).

Since the coffee trade fits squarely in the market sphere, people tend to act in it through their 'disjoint' identity, as rational choosers. In this mentality, nobody in their right mind would take the sort of risks that were necessary to get YA' up and running, especially knowing that the whole point of the effort was to enable 'free riders', that is, to improve conditions for all producers. However, YA' was able to find people willing to make this personal sacrifice for the sake of the long-term common good precisely by recruiting initial cooperative members from among the *cargos*: it appealed to their public 'conjoint' identity, mobilising its logic of community solidarity within the sphere of the market.

Agroecology, though, is the key ingredient that helps to dynamically articulate the two spheres. Most directly, promoting agroecology is a strategy for improving the ecological conditions in which Tseltal producers grow their food and their coffee. Various techniques are used, for example, to reverse the process of soil erosion, enhance soil fertility, and optimise the retention of groundwater (cf. Altieri, 1995). Caring for the soil and multi-cropping also make crops less susceptible to disease, and help to manage insects without the need for pesticides (Altieri and Nicholls, 2003). This investment in the 'base' will serve the community for generations to come.

Agroecology has further economic implications. The most obvious one is that agroecology recycles nutrients and other factors of production as much as possible, rather than relying on external inputs (and producing waste), as in the linear industrial model. This makes agricultural production relatively, if not entirely, independent of the need for money. Increased productivity depends primarily on labour and knowledge, rather than capital. Moreover, it is extremely productive in terms of scope (that is, if one considers the wide range of useful products and services that it produces), but less productive than industrial agriculture when considering only short-term yields of a single crop, and when ignoring the cost of negative externalities and subsidies from the base; it is also more resilient to shocks (Altieri et al., 2012).

All of the above factors make agroecology more conducive to middle peasant smallholding than to capitalist agriculture, in which only commodities are of interest and economies of scale tend to favour larger landholdings (Borras et al., 2015). In this way, by promoting agroecology, the MB reinforces not only ecological enhancement and food sovereignty, but also the equitable distribution of land; in turn this helps to maintain social cohesion.

The fact that agroecology is knowledge-intensive (i.e. it requires highly skilled labour) has more than economic repercussions. The specialised scientific and technical knowledge it entails is part and parcel of Tseltal culture. The affinity between agroecology and Tseltal cultural reproduction is especially clear considering the integral role of *milpa* (itself a paradigmatic form of agroecology) in traditional Tseltal life, as described above.

Cultural reproduction, social cohesion, and a 'middle peasant' model that affords a high degree of self-sufficiency (and therefore of freedom with one's time) all contribute to the viability of the *cargo* system, which is based on unpaid work at the service of the community. In turn, the *cargo* system itself is designed to strengthen the elements of Tseltal culture that foster agroecology. This is true of the specific *cargos* with a straightforward connection to the practice, the caretakers of Mother Earth (who promote agroecology), the traditional healers (who use plant-based medicine), and the promoters of *buen vivir* (who implement sustainable development projects). Beyond that, it is also true of the whole ethos of the *cargo* system itself, with its revival of the Tseltal language, ways of learning, cosmology, and moral economy.

The mutual compatibility of agroecology and the *cargo* system as such thus reinforces social cohesion on many interconnected levels. For example, the MB's *cargo*-based justice system not only promotes reconciliation, but also eschews cash fines: if there is restitution to be made, it might be in the form of chicken stew. This is highly significant in part because it fits well with the peasant economy.

Yet another social institution that simultaneously reinforces cohesion and equity, while depending on them for its own resilience, is the cooperative business through which the MB engages the coffee trade, YA'. As it turns out, Yomol A'tel's business model also has its own elective affinity with agroecology. Through upgrading strategies that enable a crucial market incentive for agroecological production, YA' reinforces the MB's wider holistic model.

Together with forward vertical integration (i.e. controlling the whole value chain), Yomol A'tel's other key strategy is product quality upgrading. By circumventing the mainstream coffee trade, which pays by weight for coffee that is not fully processed, forward vertical integration allows YA' to pay producers a price premium based on quality. Quality upgrading is important because it positions YA' in a high-value niche where there is less competition. It also helps producers overcome alienation while creating a worthy export product that represents the Tseltal people to the outside world. Just as importantly, coffee quality upgrading is achieved precisely through agroecological practices.[4]

At the production and harvest stage, two main variables improve the quality of a given coffee tree's yield: achieving high bean density and picking the cherries when they are optimally ripe.[5] High bean density is important in the specialty market because it gives the coffee its special flavour characteristics. Furthermore, it increases the efficiency of the harvest and early processing, in terms of the ratio between the total weight of the cherry (including the bean and the flesh that surrounds it) and the weight of its commodifiable part (i.e. the bean itself).[6]

There are three ways to increase density: planting at higher altitude, because it slows down the tree's metabolism; shade, because it slows down the rate of photosynthesis; and organic fertiliser, because it guarantees a balanced diet of micro- and macronutrients ingested at a natural rate, as opposed to the artificial fertilisers that are designed to pump the bean exclusively with macronutrients. (The difference is analogous to that between a free range chicken, whose meat is leaner but 'gamier', and an industrially farmed chicken, plump but insipid.) Of those three variables, altitude is hard to change, but shade and organic fertiliser are proper to agroecology.

As for picking cherries when they are optimally ripe (as opposed to the conventional practice of 'milking' branches of all their cherries at once), this requires careful labour and frequent visits to the trees. On the other hand, it spaces out the demand for labour for this specific task, because the cherries ripen at uneven times. The consequence is that optimal picking requires skilled and careful labour, but, because the knowledge and skills involved are transferable to production for use, this labour can also be combined with other tasks on the same site. Again, there is an elective affinity with agroecology, both in the type of labour it demands and in the fact that multi-cropping, with an eye to economies of scope, can make the most efficient use of this labour (Altieri et al., 2012).

Moreover, Yomol A'tel's strategy for quality upgrading reinforces the knowledge and labour-intensive aspects of agroecology that make it difficult to expand beyond a small scale. Since the business model simultaneously guarantees a high price (through this and other upgrading strategies), it makes the small scale viable. By fostering small-scale coffee plots, YA' enables families to keep designating other parts of their land for *milpa* and primary forest, which translates into a patchwork landscape of different types of land use.

This is ideal for biodiversity, not least because it creates an archipelago of habitats that migratory species, for whom Chiapas is a corridor between North and South America, need (Perfecto et al., 2009). Enhanced biodiversity and the absence of pesticides not only allows honeybees to thrive (Nicholls and Altieri, 2013), it is also ideal for high-quality honey production, which is part of Yomol A'tel's intersectoral upgrading strategy. Again, given Yomol A'tel's overall business model, there is a further market incentive for agroecology.

By capitalising on the economic incentive for a style of production that maximises biodiversity, uses sustainable practices, and minimises competition with production for use, the MB avoids a conflict between its market-engaging strategy and its base-protecting strategy, even as it creates the conditions for long-term productivity and resilience. By doing this in a way that requires immersion in the Tseltal people's culture, lifestyle, and ecosystem, the agroecological production of coffee contributes to the base in many more ways than generating income for its market provisioning. In short, agroecology helps to hold the MB's dual economic model together.

The implications of the MB's model are even clearer when contrasted with those of the industrial, input-dependent, shadeless monoculture that characterises coffee production in other parts of Chiapas. Conventional plantations produce mediocre coffee that ripens all at once, and therefore must be harvested within a narrow window of time. It then requires a high volume of short-term, unskilled (and thus low-paid, alienated) labour (Rus, 2003). It yields no other benefit but coffee, and requires more capital to produce it (in terms of agrochemicals and also because trees treated this way have shorter life expectancies, and must be replaced more frequently). Meanwhile, it causes significant ecological damage: biodiversity loss, soil erosion, and greenhouse gas emissions from fertiliser production and transport, which contribute to climate change. This debasement reduces the long-term viability of production. Finally, this model will tend to increase inequality, because it lends itself to economies of scale even as it deteriorates food sovereignty and cultural reproduction, thereby weakening social cohesion (cf. Thorp, 1998).

9. Implications for wider society

The MB model also has direct and indirect implications beyond northern Chiapas. Yomol A'tel's coffee shops, rooted in the same social economy ethos as the rest of the group, have created decent employment opportunities for residents of Mexico City, Puebla, and Guadalajara. The shops, which are largely based on university campuses, have also served as a platform to engage urban consumers around social, ecological, and economic issues in parts of their country that are otherwise largely invisible to them. An even stronger rural-urban link, entailing a financial cooperative whose members would include producers, workers, and consumers, is currently in the design stage.

More indirectly, but no less importantly, the MB's dual economy contributes to the wider common good (cf. Riordan and Travieso, 2018). By supplying domestically roasted coffee to cities, YA' contributes to Mexican food sovereignty. Also, by enabling Tseltal peasants to continue making *milpa*, the MB fosters the in situ conservation of maize biodiversity, thereby bolstering worldwide food security (Isakson, 2009). There are, of course, many other positive externalities of agroecology. However, perhaps the farthest reaching way in which the MB helps us all is by showing us that it is indeed possible to have healthy, sustainable, and socially just food systems, as part of a viable, life-giving post-capitalist economy.

10. Survival and impact in an adverse system

Despite its obvious advantages, the MB's model is up against an overarching system (including laws, institutions, research paradigms, etc.) that is still largely designed at the service of capitalism. In addition to the illustrations of this in the discussion above, Mexican legislation has made it increasingly difficult to operate cooperative financial institutions (Oulhaj, 2016), and the government has subsidised the efforts of multinational corporations to promote shadeless, low-quality coffee in Tseltal territory.

To survive in the midst of an adverse system, the MB cannot go it alone. It actively participates in various networks, especially throughout Mexico and Latin America. The MB joins its peers in the indigenous agroecology, food sovereignty, and social and solidarity economy movements. It also makes cross-sectoral alliances, for example with universities. In these collaborations, the MB finds both a protected space in which to mature its model and a channel for wider impact. Through the Comparte network, for example, some of the MB's innovations are already being replicated in other countries.

11. Conclusions

Going back to our allegory of the lion, the MB's creative articulation of both solidarity economy and social economy strategies allows it to overcome the risks of each (isolation leading to unviability on the one hand, and inadvertent commodification leading to debasement on the other). Further, it generates positive feedback loops that transfer value between spheres, creating new economic, social, and ecological value in the process. By re-embedding food systems in our relationships with nature and with each other, the model sets in motion dynamics of resourcefulness. Whereas 'resilience' describes the ability to withstand external shocks, 'resourcefulness' implies the ability to thrive by creatively engaging the sources of those shocks (MacKinnon and Derickson, 2012).

However, even in the best-case scenario of successful systemic transformation, the MB model entails a permanent built-in tension. As we have seen, the MB's model depends on first separating and then conjugating the realm of mutuality (community and base) on the one hand, and the realm of the market (mainly through coffee) on the other. We have also noted that these two spheres are at once mutually interdependent and mutually contradictory.

There is a real tension between the two dimensions of the MB economy. For example, the community and its ecological base have a circular, slow-paced rhythm, while engaging the market requires linear acceleration (cf. Bear, 2017). Community and market also imply different modes of reason and ways of relating.

The challenge, then, is to make the tension a creative one. To this end, the MB applies the core Tseltal value of 'harmony', whereby diverse notes make a beautiful sound, greater than the sum of its parts. In this chapter I have tried to show how agroecology, food sovereignty, social economy, and solidarity economy interact to harmonise the spheres of the MB's economy, creating possibilities that reverberate into broader debates.

Although it is unique in many ways, the logic and implications of the MB's economic design can contribute to the ongoing search for new possibilities in sustainable food planning. Its use of agroecology is essential to its complex vision, in ways that connect rural indigenous food sovereignty to small-town jobs to sociopolitical engagement with big-city consumers. Indeed, scholars have begun to take note of the model's sophistication,[7] and grassroots movements are emulating its innovations. Surely there is room for further research and experimentation.

Notes

1 This section and the next include some paragraphs that are translated and adapted from my own work in a Spanish-language publication (Travieso, 2016). The translation is used here with permission.
2 Gudeman is recognised by his peers as having carried out the most systematic effort to build a general theory of economy (cf. Gregory, 2002; Escobar, 2008: 73; Hann and Hart, 2011: 84; Hart, 2017). Many readers will recognise the partial affinity with Polanyi's seminal work; for more on that, cf. Gudeman, 2009.
3 The name means 'we work together' in Tseltal.
4 The following discussion on coffee and agroecology is based primarily on my interview with the agroecologist, José Alejandro Aguilar. He completed his doctorate under Miguel Altieri, has worked with peasant communities in Colombia for decades, and now runs an experimental research farm that produces prize-winning specialty coffee. In addition to these credentials, he is a close advisor to YA'; his understanding has largely informed Yomol A'tel's own vision and practice.
5 The variety of coffee is also a factor in market-recognised quality. Generally speaking, higher quality varieties (e.g. arabica) tend to prefer shade, while lower quality varieties (e.g. robusta) need more sunlight and can be planted densely together as monoculture. Again, agroecology, multi-cropping, and quality go together.
6 The coffee cherry's pulp and the hull are not useful to buyers (producers can use them as organic fertiliser); only the bean is of commercial interest. Therefore, the density of beans is a factor in the cost-efficiency of harvest and wet and dry processing, as well as transportation. The denser the bean, the more cost-effective the process. To be sure, most buyers do not reward producers for this, just as they do not incentivise quality. Yomol A'tel's ability to recognise this factor again depends on its having achieved forward vertical integration. In other words, YA' has designed its circuit to make the most of this link between market value and agroecology.
7 For an extensive bibliography, see https://yomolnohptesel.wordpress.com/biblioteca/ya-mb/.

References

Altieri, M. A. (1995) *Agroecology. The Science of Sustainable Agriculture*. Boulder: Westview.

Altieri, M. A., Funes-Monzote, F. R. and Petersen, P. (2012) 'Agroecologically efficient agricultural systems for smallholder farmers: Contributions to food sovereignty', *Agronomy for Sustainable Development*, vol 32, no 1, pp1–13.

Altieri, M. A. and Nicholls, C. I. (2003) 'Soil fertility management and insect pests: Harmonizing soil and plant health in agroecosystems', *Soil and Tillage Research*, vol 72, no 2, pp203–211.

Altieri, M. A. and Toledo, V. M. (2011) 'The agroecological revolution in Latin America: Rescuing nature, ensuring food sovereignty and empowering peasants', *Journal of Peasant Studies*, vol 38, no 3, pp587–612.

Barnes, G. (2009) 'The evolution and resilience of community-based land tenure in rural Mexico', *Land Use Policy*, vol 26, no 2, pp393–400.

Barrientos, S., Gereffi, G. and Rossi, A. (2011) 'Economic and social upgrading in global production networks: A new paradigm for a changing world', *International Labour Review*, vol 150, nos 3–4, pp319–340.

Bear, L. (2017) 'Anthropological futures: A critical political economy of capitalist time: ASA's Raymond Firth lecture, 2016', *Social Anthropology*, vol 25, no 2, pp142–158.

Bobrow-Strain, A. (2007) *Intimate Enemies. Landowners, Power and Violence in Chiapas*. London: Duke UP.

Borras Jr, S. M., Franco, J. C. and Monsalve Suárez, S. (2015) 'Land and food sovereignty', *Third World Quarterly*, vol 36, no 3, pp600–617.

Carsten, J. (1989) 'Cooking money: Gender and the symbolic transformation of means of exchange in a Malay fishing community', in J. Parry and M. Bloch (eds) *Money and the Morality of Exchange*. Cambridge: Cambridge UP, pp117–141.

Comberti, C., Thornton, T. F., de Echeverria, V. W. and Patterson, T. (2015) 'Ecosystem services or services to ecosystems? Valuing cultivation and reciprocal relationships between humans and ecosystems', *Global Environmental Change*, vol 34, pp247–262.

CONEVAL, Consejo Nacional de Evaluación de la Política de Desarrollo Social (2012) 'Mapa interactivo de pobreza por entidades federativas'. Available at: http://www.coneval.gob.mx/Medicion/Paginas/Medici%C3%B3n/Pobreza%202012/Mapa-interactivo-de-pobreza-por-entidades-federativas-2012.aspx.

Crispín Bernardo, M. and Ruiz Muñoz, M. (2010) *Huellas de un caminar. La Misión jesuita de Bachajón*. México: Universidad Iberoamericana.

Dacheux, Éric (2013) 'Les logiques plurielles des mouvements économiques anticapitalistes', in B. Frère and M. Jacquemain (eds) *Résister au Quotidien?* Paris: Presses de SciencesPo, pp111–132.

Daigle, M. (2019) 'Tracing the terrain of Indigenous food sovereignties', *Journal of Peasant Studies*, vol 46, no 2, pp297–315.

Daviron, B. and Ponte, S. (2005) *The Coffee Paradox: Global Markets, Commodity Trade and the Elusive Promise of Development*. London: Zed.

De Schutter, O. (2010) *Promotion and Protection of All Human Rights, Civil, Political, Economic, Social and Cultural Rights, Including the Right to Develop*, Report submitted by the Special Rapporteur on the right to food: Human Rights Council 16th Session: Agenda Item 3. UN General Assembly. A/HRC/16/49.

De Schutter, O. (2014) 'The Specter of Productivism and Food Democracy', *Wisconsin Law Review*, vol 2014, no 2, pp199–233.

Escobar, A. (2008) *Territories of Difference: Place, Movements, Life, Redes*. Durham: Duke UP.

FAO, IFAD and WFP (2015) *The State of Food Insecurity in the World 2015. Meeting the 2015 International Hunger Targets: Taking Stock of Uneven Progress*. Rome: FAO.

Francis, C., Lieblin, G., Gliessman, S., Breland, T. A., Creamer, N., Harwood, R., Salomonsson, L., Helenius, J., Rickerl, D., Salvador, R., Wiedenhoeft, M., Simmons, S., Allen, P., Altieri, M., Flora, C. and Poincelot, R. (2003) 'Agroecology: The ecology of food systems', *Journal of Sustainable Agriculture*, vol 22, no 3, pp99–118.

González de Molina, M. G. (2013) 'Agroecology and politics. How to get Sustainability? About the necessity for a political agroecology', *Agroecology and Sustainable Food Systems*, vol 37, no 1, pp45–59.

Gregory, C. A. (1997) *Savage Money: The Anthropology and Politics of Commodity Exchange*. Amsterdam: Harwood Academic.

Gregory, C. A. (2002) 'Review of Gudeman (2001): The anthropology of economy', *Australian Journal of Anthropology*, vol 13, no 3, pp361–362.

Gudeman, S. (2008) *Economy's Tension. The Dialectics of Community and Market*. Oxford: Berghahn.

Gudeman, S. (2009) 'Necessity or contingency: Mutuality and the market', in C. Hann and K. Hart (eds) *Market and Society: The Great Transformation Today*. Cambridge: Cambridge UP, pp17–37.

Gudeman, S. (2016) *Anthropology and Economy*. Cambridge: Cambridge UP.

Hann, C. and Hart, K. (2011) *Economic Anthropology: History, Ethnography, Critique*. Cambridge: Polity.

Hart, K. (2017) 'Anthropology and the economy, by Stephen Gudeman', *Anthropological Forum*, vol 27, no 3, pp287–289.

Hart, K., Laville, J. L. and Cattani, A. D. (2010) *The Human Economy*. Cambridge: Polity.

Hecht, S. (1995) 'The evolution of agroecological thought', in M. A. Altieri (ed.) *Agroecology. The Science of Sustainable Agriculture*. Boulder: Westview, pp1–20.

Hoffman, U. (2013) 'Agriculture at a crossroads: Assuring food security in developing countries under the challenges of global warming', in UNCTAD (ed.) *Trade and Environment Review 2013. Wake Up Before It Is Too Late*. Geneva: UNCTAD, pp2–8.

Holt-Giménez, E. and Altieri, M. A. (2013) 'Agroecology, food sovereignty and the new green revolution', *Agroecology and Sustainable Food Systems*, vol 37, no 1, pp90–102.

INEGI, Instituto Nacional de Estadística y Geografía (2010) *Panorama Sociodemográfico de México*. Mexico City.

Irezabal Vilaclara, A. (2015) 'La empresa social y solidaria para la defensa del territorio: Una experiencia en Chiapas, Mexico', in I. Alzola, R. Altuna, S. Arando (eds.) *La Economía Social, Cuestiones de Actualidad*. Mondragon: Mondragon Unibertsitateko Zerbitzu Editoriala, pp65–93.

Isakson, S. R. (2009) 'No hay ganancia en la milpa: The agrarian question, food sovereignty, and the on-farm conservation of agrobiodiversity in Guatemalan highlands', *Journal of Peasant Studies*, vol 36, no 4, pp725–59.

Jones, A., Pimbert, M. and Jiggens, J. (2011) *Virtuous Circles: Values, Systems and Sustainability*. London: IIED and IUCN CEESP. Available at: https://pubs.iied.org/14619IIED/?k=Virtuous+circles%3A+values%2C+systems+and+sustainability.

Li, T. M. (2015) 'Can there be food sovereignty here?', *The Journal of Peasant Studies*, vol 42, no 1, pp205–211.

Little, W. (2004) *Mayas in the Marketplace: Tourism, Globalization and Cultural Identity*. Austin: U Texas P.

MacKinnon, D. and Derickson, K. D. (2012) 'From resilience to resourcefulness: A critique of resilience policy and activism', *Progress in Human Geography*, vol 37, no 2, pp253–270.

McIntyre, B., Herren, H., Wakhungu, J. and Watson, R. (eds) (2009) *Agriculture at a Crossroads*. Synthesis Report. Washington: IAASTD.

McMurtry, J. J. (2013) 'Prometheus, Trojan Horse or Frankenstein? The social and solidarity economy as community creation, market wedge, or state monster', draft paper prepared for the *UNRISD Conference Potential and Limits of Social and Solidarity Economy*, 6–8 May 2013, Geneva, Switzerland. Available at: http://www.unrisd.org.

Méndez, V. E., Bacon, C. M. and Cohen, R. (2013) 'Agroecology as a transdisciplinary, participatory and action-oriented approach', *Agroecology and Sustainable Food Systems*, vol 37, no 1, pp3–18.

MRGI, Minority Rights Group International (2011) *State of the World's Minorities and Indigenous Peoples 2011 – Mexico*. Available at: http://www.refworld.org/docid/4e16d368c.html.

Nicholls, C. I. and Altieri, M. A. (2013) 'Plant biodiversity enhances bees and other insect pollinators in agroecosystems. A review', *Agronomy for Sustainable Development*, vol 33, no 2, pp257–274.

Oulhaj, L. (2016) 'Reflexión en torno a los retos de la inclusión financiera', in L. Oulhaj, (ed.) *Avanzar en la inclusión financiera*. Mexico City: Universidad Iberoamericana, pp187–205.

Parry, J. and Bloch, M. (1989) 'Introduction: Money and the morality of exchange", in J. Parry and M. Bloch (eds) *Money and the Morality of Exchange*. Cambridge: Cambridge UP, pp1–32.

Perfecto, I., Vandermeer, J. and Wright, A. (2009) *Nature's Matrix: Linking Agriculture, Conservation and Food Sovereignty*. London: Earthscan.

Reinert, E. (2003) 'Increasing poverty in a globalised world: Marshall plans and Morgenthau plans as mechanisms of polarisation of world incomes', in H. J. Chang (ed) *Rethinking Development Economics*. London: Anthem, pp453–478.

Riordan, P. and Travieso, E. (2018) 'Integral ecology and common goods in chiapas: The Misión de Bachajón in Light of Laudato Si', *Heythrop Journal*, vol 59, no 6, pp959–968.

Rus, J. (2003) 'Coffee and the recolonization of highland Chiapas, Mexico: Indian communities and plantation labor, 1892–1912', in W. G. Clarence-Smith and S. Topik (eds) *The Global Coffee Economy in Africa, Asia, and Latin America, 1500–1989*. Cambridge: Cambridge UP, pp257–285.

Sánchez Juárez, G. K. (2015) *Los pequeños cafeticultores de Chiapas: Organización y resistencia al mercado*. Tuxtla: Universidad de Ciencias y Artes de Chiapas.

Spratt, S. (2015) 'Financing green transformations', in I. Scoones, M. Leach and P. Newell (eds) *The Politics of Green Transformations*. London and New York: Routledge, pp153–159.

Thorp, R. (1998) *Progress, Poverty and Exclusion: An Economic History of Latin America in the 20th Century*. Washington: Inter-American Development Bank.

Travieso, E. (2016) 'Creatividad ante la crisis: Alternativas desde América Latina', *Ibero*, vol 8, no 45, pp36–38.

Travieso, E., Irezabal, A. and Torres, M. (forthcoming) 'La economía social y solidaria y el derecho a (no) migrar en México', in L. Oulhaj (ed.) *Miradas con perspectiva de género a la Economía Social y Solidaria como una de las opciones en el nuevo escenario de la migración*. Mexico City: Universidad Iberoamericana.

Tsing, A. L. (2015) *The Mushroom at the End of the World: On the Possibility of Life in Capitalist Ruins*. Princeton, NJ: Princeton UP.

Van der Ploeg, J. D. (2010) 'The food crisis, industrialized farming and the imperial regime', *Journal of Agrarian Change*, vol 10, no 1, pp98–106.

Vogt, E. Z. (1969) *Zinacantan: A Maya Community in the Highlands of Chiapas*. Cambridge: Harvard UP.

Wilson, J. (2011) 'Notes on the rural city: Henry Lefebvre and the transformation of everyday life in Chiapas, Mexico', *Environment and Planning D: Society and Space*, vol 29, no 6, pp993–1009.

Zatyrka, A. (2003) *The Formation of the Autochthonous Church and the Inculturation of Christian Ministries in the Indian Cultures of America, a Case Study: The Bachajón Mission and the Diocese of San Cristóbal de las Casas, Mexico*. Doctoral dissertation in Theology (Leopold Franzens Universität, Innsbruck, Austria).

10

PEASANT COUNTER-HEGEMONY TOWARDS POST-CAPITALIST FOOD SOVEREIGNTY

Facing rural and urban precarity

Mark Tilzey

Introduction

This chapter argues that there are compelling reasons, both social and biophysical, to sustain advocacy of a counter-hegemonic, anti-capitalist, and radical position to which the notion of food sovereignty is central. Such advocacy addresses the structural foundations of capital's twin 'political' and 'ecological' contradictions (Tilzey, 2018); these derive from the commodification of labour power attendant on the expropriation of producers from their means of production – the process of "primitive accumulation" (Marx, 1972). Such expropriation leads to a condition of "market dependency" (Kautsky, 1988; Wood, 2002) in order to secure the means of livelihood. 'Radical' food sovereignty envisages an abrogation of this condition as a prerequisite for social and ecological sustainability, a "political (agro)ecology of praxis" (Tilzey, 2018). This praxis includes social movements advocating urban agroecology, through which they hope to confront and subvert the 'neoliberalisation of the city' by addressing 'food precarity' on the basis of non-capitalist/market provisioning, such as 'commoning'. The purpose of this chapter is to provide a broader spatio-temporal contextualisation of such urban agroecological movements by placing them within the wider 'political' and 'ecological' dynamics of the state-capital nexus, and of the imperial, sub-imperial, and peripheral form of the world capitalist system that structures these. The rationale behind this is to afford a strategic view on the opportunities and constraints surrounding the 'out' and 'upscaling' of urban agroecological movements. In short, this chapter seeks to present a "strategic relational" view (Jessop, 2005) of the state/capital/class parameters that underpin the 'political' and 'ecological' resourcing of an agroecological urbanism.

At the outset, we argue, this requires a more precise definition of capitalism and anti-capitalism, and a better understanding of capital-state dynamics such as imperialism (and now sub-imperialism) (Tilzey, 2016) if market dependency and primitive accumulation are to be subverted through a 'new frontier' of resistance as counter-hegemony (see Tilzey, 2017b). It is argued here that this new frontier of resistance, as 'radical' food sovereignty, and for reasons intimately related to the imperialistic and sub-imperialistic character of capitalism as the "imperial mode of living" (Brand and Wissen, 2018), is differentially located in the Global South. This is the global locus of the majority of peasantry, semi-proletarians, and indigenous people whose continuing links to non-commodified land offer escape routes from the precarity of 'disarticulated' capitalism ('jobless growth' or 'growth with precarity').

We argue that this differential location occurs because the capitalist 'agrarian transition' in the Global South has not generally taken the form of the full proletarianisation (complete separation of workers from land as the basic means of production) that has usually characterised the Global North and parts of the sub-imperium (notably China). Rather, the 'transition' is often partial, with peasants, as semi-proletarians, commonly retaining varying degrees of access to land, a tendency compounded by the lack of secure employment opportunities in capitalist agriculture and in the non-agrarian sector. This is due, to a significant degree, to peripheral and dependent forms of capital accumulation. Due to semi-proletarianisation and precarity, we argue that a 'radical' imaginary of food sovereignty tends to emerge, a counter-hegemonic position that advocates the redistribution of land from capitalists to the peasantry and precariat for the production of fundamental use values, perhaps most importantly food, often employing agroecological principles. Such a 'reverse' agrarian transition, focusing on access to land in the countryside as much as in the city, problematises the capitalist dualism of city and countryside, and seeks to subvert the 'metabolic rift' that sustains this division.

We suggest that the capitalist 'agrarian transition' in the Global North (and increasingly in the sub-imperium) has, by contrast, entailed proletarianisation, the full separation of workers from the means of production on the land, leading to the extirpation of the peasantry and the almost exclusive appropriation of the countryside by capitalist farmers. The potentially revolutionary character of the proletariat here has typically been dulled by the co-optation of this class into capitalism as 'labour aristocracy' and as 'consumers'. Such 'material rewards', and the prevalently urban character of these societies, have been facilitated by 'unequal ecological exchange' with the Global South, something we may term 'resource imperialism'. The implication of these dynamics is that, due to the general extirpation of the peasantry, rural resistance to capitalism in the Global North is miniscule by comparison to the South, and commonly assumes the form, not of 'radical', but rather of 'progressive', food sovereignty as advocated by 'alternative', commercial, smaller scale producers. By the same token, more radical movements in the Global North tend to have greater association with the city, the primary locus of the newly emergent precariat of neoliberalism, taking the form of urban agroecology and calls for the 'right to the city'. The corollary of this, we argue, is that the 'right to the countryside', a countryside now occupied almost exclusively by capitalist farming, has been relatively neglected, with this neglect sustained to a significant degree by the 'imperial mode of living'.

We suggest that urban agroecology, as food sovereignty, constitutes an important 'beachhead' against capitalism by addressing precarity through counter-hegemonic modes of local food provisioning. We also argue, however, that if this 'beachhead' is to be secured and widened, the 'right to the city' needs to go hand in hand with calls for the 'right to the countryside' as part of a broader counter-hegemonic movement that seeks to 'visibilise' the 'imperial mode of living' and to institute ecological sustainability and food equity as key components of this. In the Global North, this will not be easy, however.

It is here, we suggest, that the role of the Southern precariat becomes pivotal. This is not merely through example. It is also as potential agent in fomenting transitions to post-capitalism beyond the South by 'closing down' resource imperialism (otherwise known as 'neo-extractivism'), and thereby disabling the capacity of the Northern state-capital nexus to co-opt counter-hegemony through the 'imperial mode of living'. The North will be increasingly obliged to fall back on its own resources, stimulating the growing precariat to seek direct access to the means of production as land in both the city and, perhaps more particularly, in the countryside. With enforced 'energy descent' and 'de-growth' contingent on the termination of resource imperialism, it will become an increasing challenge for the state-capital nexus to perpetuate the 'metabolic rift' between countryside and city.

We speculate that, at this point, the demand by an ever-enlarging precariat in the Global North for access to land will ramify outwards from the city to encompass the countryside. Enforced energy descent and de-growth will compel, for ecological reasons, the general adoption of agroecology, while the demand for food equity on the part of the precariat, particularly, will foment the spread of 'radical' food sovereignty. No inevitability is implied here, of course, since any such scenario is contingent upon 'political' dynamics (the relative strength and strategic dexterity of social forces, or 'class struggle'), albeit against an 'ecological' backdrop of ineluctably increasing resource and energy scarcity. Indeed, the difficulties entailed in subverting the state-capital nexus, even in the Global South (let alone the Global North), are demonstrated in this chapter by reference to the experience of counter-hegemonic movements in the 'pink tide' states of Latin America.

The struggle against primitive accumulation and market dependency through 'radical' food sovereignty

Marx (1972) recognised that "primitive accumulation" was pivotal to the birth of capitalism, entailing, simultaneously, the expropriation of peasant producers, the creation of a labour force 'free' to sell its labour power to the capitalist, the commodification of land through the conferral of absolute property rights, and the growth of cities and industry through the new relationship between capital and labour. Primitive accumulation took place in its 'pristine' form with the enclosure movement of 17th-, 18th-, and 19th-century Britain; it was subsequently extended, internationally, to encompass much of the globe through colonisation and imperialism. The peasantry and indigenous people were thereby expropriated, wholly or partially, from their direct and customary access to the means of subsistence, a process that continues today and is accelerating in the Global South, particularly, as remaining non-commodified lands are subsumed within capital. As we explain further below, the necessarily state-dependent and imperialistic nature of capitalism means that, in its 'periphery' (the Global South), agrarian 'transition' may not take the form of the generally complete process of proletarianisation, and the formation of 'permanent' cities, that has generally characterised the 'core' states of the Global North and parts of the sub-imperium. Rather, this often assumes the form of partial transition, or semi-proletarianisation, in which peasantries retain variable degrees of access to land. This is due to both peripheral forms of capital accumulation, including non-capitalist forms of production, and *resistances* to them on the part of subaltern classes.

It is unsurprising, therefore, that primitive accumulation has been associated historically with profound contestations surrounding access to land as the basis of the means of production. Today, as capitalism in its various guises continues the process of enclosure, dispossession, and (semi)proletarianisation (but under conditions of 'jobless growth', 'growth with precarity', and increasing ecological contradiction), struggles over the conditions of livelihood are increasing. A key part of the legitimating narrative of capitalism is the portrayal of this process of agrarian transition to modernism as an apolitical and ineluctable teleology, supported by an imaginary involving the conversion of former peasants into workers with relatively secure prospects of employment, usually through migration to urban-based industry (Perelman, 2000). Such an imaginary is no longer tenable throughout much of the Global South (and increasingly outside the periphery). The reality of precarity, involving poor pay, intermittent employment, or self-exploitation in the informal sector, erodes this narrative of the inevitability and superiority of the capitalist/urban/industrial mode of existence vis-à-vis non-capitalist modes. The result is a 'radical' imaginary of food sovereignty, indeed of land sovereignty or livelihood sovereignty (Tilzey, 2018), as counter-hegemony and as a *reverse* agrarian transition, in which access to land, both in the city and, particularly, in the countryside, for production of essential use values, is the foremost demand of the precariat. By implication, the 'radical' food sovereignty counter-narrative is one of obstructing and reversing primitive accumulation and capitalist market dependence through the reappropriation of land and resources in both countryside and city as the basis for socially equitable, cooperative, and ecologically sustainable production, primarily of food, for family and community. This is to subvert the metabolic rift, envisioning *rus in urbe* and *urbs in rure*.

Capitalism, imperialism, and the imperial mode of living

Through hegemony, involving the discursive and material co-option of non-capitalist classes, and within the necessary context of the state-capital nexus, capitalism has been remarkably successful in neutralising resistance to its exploitative dynamic. This success, however, has been located differentially in the Global North, and now to some degree in the sub-imperium, notably China. An essential part of this ability to neutralise and co-opt derives from the capitalist world system's broadly bipolar (possibly tri-polar) structure: the socially 'articulated' states of the Global North and the socially 'disarticulated' states of the Global South. Social articulation implies a complementarity between the role of the labour force as producers and consumers and a situation in which its role as consumer outweighs its significance as producer. This entails the co-optation of non-capitalist classes as 'consumers' into the capitalist "norm complex" (Bernstein, 2002). However, as the 'imperial mode of living' (see below), such hegemony is possible only through imperialistic relationships with the Global South, from whence surplus value is appropriated to feed generalised consumerism in the Global North (and the sub-imperium). Tendentially, therefore, oppositional relationships between capitalist and non-capitalist classes in 'articulated' states have been defused by 'flanking' measures based on material and discursive rewards based around redistributive, social welfare, nation-building, environmental, and other policies, together with the bestowal of 'liberal' citizenship rights (Moyo and Yeros, 2011; Chibber, 2013; Mooers, 2014).

In contrast, social disarticulation occurs when the state-capital nexus is interested in its labour force principally from the perspective of production (its ability to generate surplus value) and not primarily from the perspective of consumption (the realisation of surplus value through the sale of commodities). In other words, the labour force is significant as a source of surplus value, but not as a source of consumption. Typically, therefore, the global Southern state-capital nexus has only weakly developed 'flanking' measures, or modes of regulation, through which to embed capitalism materially and ideologically. Its producers, the subaltern classes, are only weakly incorporated as consumer beneficiaries, therefore, and the state has low legitimacy (Tilzey, 2018).

This 'real' subsumption of non-capitalist classes as a 'labour aristocracy' in the Global North reached its height during the post-war Keynesian era. In the face of increased class exploitation since the advent of the neoliberal era in the 1980s, attempts to sustain this compact between capital and non-capitalist classes in the Global North have been undertaken increasingly by means of (renewed) imperial relations with the Global South. Surplus value from the classes of labour now flows from South to North, 'subsidised' by the massive and destructive haemorrhage of 'ecological surplus' that lies behind this relationship (Exner et al., 2013; Moore, 2015; Smith, 2016). Burgeoning levels of social and ecological dislocation in the South have been the consequence of this neo-extractive relationship. This has been, and continues to be, associated with resurgent neoliberal primitive accumulation, undermining the incipient legitimacy, employment, and welfare functions that had been, at least in some degree, undertaken by the 'peripheral' state until the 1980s. The outcome of this 'new imperial' relationship between North and South (Harvey, 2003; Smith, 2016) is that the citizens of the former, despite certain attenuation, continue to be accorded privileges, material, welfare, legal, etc., denied to those in the capitalist periphery (see, for example, Mooers, 2014). The ability of this exploitative relationship between Global North and Global South to support generalised consumerism (social articulation) may be termed the *imperial mode of regulation*, or the *imperial mode of living* (Brand and Wissen, 2012, 2018).

The imperial mode of living of the Global North (and selectively of the sub-imperium) is sustained by the state-capital nexus on the basis of the unlimited appropriation of resources and labour power from the Global South, and on a disproportionate claim upon global ecological sinks, that is, the capacity of the environment to absorb waste. In the multiple crises of late capitalism – economic, social, ecological – the imperial mode of living helps to sustain social stability in the Global North through a continuing supply of consumer 'rewards', whilst affording a hegemonic orientation (the aspiration to consume) on the part of those not yet on, or on the lower rungs of, the capitalist 'career ladder'. It thus refers to the dominant patterns of capitalist production, distribution, and consumption that are deeply rooted in the everyday practices of the majority in the Global North, and increasing in the sub-imperium, most notably China. It also comprises an aspiration for many millions of the global precariat, but one that cannot be realised both for 'political' and 'ecological' reasons (see Tilzey, 2018). The imperial mode of living thus explains the relative political and social quiescence of workforces in the Global North in the face of increasing global turbulence, whilst at the same time explaining the very basis of that turbulence in terms of capital's unlimited call upon surplus labour, and the 'ecological surplus' lying behind this (Tilzey, 2018), now differentially located in the Global South.

The development of 'social articulation' in the Global North and the sub-imperium is thus premised on a world resource system hugely biased in favour of these centres of accumulation. Where Fordism was based on a form of intensive accumulation associated with an increase in relative surplus value secured through permanent intensification of the labour process and expansion in labour productivity (based in turn on the 'ecological surplus' arising through the intensified exploitation of fossil fuels), post-Fordism, since the 1980s and particularly since the 1990s, has witnessed a refocus on *extensive* forms of accumulation. This involves a partial reorientation, in combination with relative surplus value, towards *absolute* surplus value such as the flexibilisation and expansion of the working day, and the suppression of wages *below* the reproductive needs of the workforce, increasingly a precariat (Smith, 2016; Tilzey, 2018). It is the Global South, however, that has borne the overwhelming brunt of this trend, while the North has differentially continued to focus on relative surplus value generation in 'high value-added' industries. This restructuring of the international division of labour has succeeded in intensifying imperial access, not only to the labour power capacities of global Southern states, but also to their resources. Here labour power, 'super-exploited' through the realisation of absolute surplus value extraction, is routinely subsidised by non-commodified (unremunerated) family reproductive activities, often undertaken by women, and by the unsustainable utilisation of the environment. The perpetuation of a *semi-proletariat*, with partial but inadequate access to land and in functional dualistic relation to capital, is key to this "new imperial" (Harvey, 2003; Tilzey, 2016) subordination of the South to the North.

Thus, for example, while the consumption by the European Union of resources from within its own territory has been stagnant at a fairly high level since the mid-1980s (when post-Fordism/neoliberalism took hold), its share of imported resources from the Global South has increased markedly, together with the "ecological backpack" embodied in those imports (Brand and Wissen, 2012). In addition to the extraction from the Global South of surplus labour and the 'ecological surplus' underlying this, the third dimension of the imperial mode of living is the disproportionate claim of the Global North and sub-imperium on the Earth's sinks, generating the overall background turbulence manifest in 'capitalogenic' climate change, the adverse impacts of which are felt differentially by those most vulnerable and least culpable – the subaltern classes of the Global South. Thus, not only is the relationship between North and South characterised by the 'political' super-exploitation of labour (Smith, 2016), it is also characterised by unequal 'ecological' exchange.

An important aspect of the imperial mode of living, therefore, is the reproduction of the workforce as 'labour aristocracy', the complex division of labour associated with post-agrarian and urban-dominated capitalism and the hegemonic legitimation of this material and ideological constellation. This constellation is predicated on access to cheap raw materials and labour power, overwhelmingly located in the Global South, together with the ability to use global ecological sinks in a disproportionate way. The imperial mode of living thus explains why neoliberal policies in the Global North either have not yet exhausted their legitimisation potential or are turning in 'authoritarian' or 'national populist' directions (Trumpism, Brexit). This is as citizens of the North demand their 'right' to continue high consumption lifestyles in the face of burgeoning threats (immigration, terrorism, climate change, etc.) induced by, but causally disassociated from, that very mode of living.

The state plays a major role in constituting and stabilising the imperial mode of living by not only securing access externally to strategic resources, but also guaranteeing internally a certain standard of living of the masses through social insurance systems and labour market regulations (even if these have suffered relative degradation under neoliberalism). The state acts to implement a mode of regulation as the hegemonic normalisation of consumer affluence as the key denominator of the 'good life', while the ability to promise and secure growth and 'progress' (the regime of accumulation) is particularly important since it constitutes the material basis of the imperial mode of living. The imperial mode of living, then, is of central relevance to this chapter because:

1. It means that post-capitalist socio-ecological transition is being hampered or blocked by powerful economic and political interest groups, by the imperial state-capital nexus more generally, and by the hegemonic status of capitalist consumerism in the Global North and sub-imperium. Therefore, such a transition is both most difficult and least likely in the Global North, precisely because of the hegemonic status of the imperial mode of living in respect to the majority of the population and its consequent unwillingness to countenance radical changes to its mode of living.
2. By contrast, in the Global South, and as a direct outcome of the operation of the imperial mode of living, the majority differentially suffers the dis-benefits of capitalist exploitation, receives few rewards either directly or via the state in terms of welfare and employment security, and is not, therefore, incorporated hegemonically into the project of consumerism. Moreover, because of the semi-proletarian status of the majority in the Global South, the material and cultural/ideological links to non-capitalist alternatives, most significantly to land, remain. In short, subaltern classes in the Global South have the least to lose and the most to gain from a transition to post-capitalism. This, therefore, gives strong indications as to which classes and class fractions are most likely to constitute the forces of change in this transition.
3. The imperial mode of living, and more generally the political ecology paradigm within which it sits, constitutes a powerful explanatory framework for both the 'political' and the 'ecological' dynamics of the multiple crises of late capitalism and their differentiated impacts. It brings to the fore the importance of the state to capital (the state-capital nexus), the importance of differentiation within the capitalist world system as 'core', 'semi-periphery', and 'periphery'[1], and the profound relevance of ecological limits to any definition of post-capitalism and the 'good life'. This means that socialism cannot be conceived as equality of affluence for all, or as a continuation of the complex divisions of labour, and the heavy urban bias, that is centrally dependent on cheap energy from fossil fuels.

The foundations of counter-hegemonic resistance to the imperial mode of living

The differential impacts in the South of capital's exploitative 'political' and 'ecological' dynamic, deficits in the 'peripheral' state's performance of its welfare, employment, and legitimacy functions, together with the 'formal' rather than 'real' subsumption within capital of the semi-proletarian majority, carry with them the increased likelihood of challenge to the state-capital nexus by counter-hegemonic social forces. Attempted reappropriations of the state, or of autonomous zones 'outside' the state, by counter-hegemonic social forces are implied, comprising resurgent national, and new post-national, forms of sovereignty. The recent re-emergence of land struggles throughout the South is indeed a response to the profound social and ecological crisis engendered by the implementation of the 'new imperial', neoliberal policies. The rural uprisings, and advocacy of 'radical' food sovereignty in Bolivia, Brazil, Ecuador, Mexico, Nepal, and elsewhere, are symptomatic of this trend (see, for example, Moyo and Yeros, 2005).

Since the previous models of quasi-Keynesian development in the South were already deficient in producing 'urban social safety valves' in the form of employment opportunities, and instead had resulted in high levels of unemployment and an enormous informal sector, the new imperialism has simply exacerbated these consequences to unprecedented levels of precarity for the majority. De Janvry et al. (1989) refer to a "double (under-)development squeeze"; this involves both a "land squeeze" as smallholders are increasingly less able to increase their landholdings or even to secure land, and an "unemployment squeeze" due to the reduction in employment opportunities both in the countryside and in urban centres. *In this context, access to land becomes a refuge from neoliberal restructuring.* The principal objective of families involved in land struggles, such as the MST (*Movimento de Trabalhadores Rurais sem Terra*) in Brazil and the Zapatistas in Mexico, is to secure, as a minimum requirement, their subsistence, or the social reproduction of the family household, through the production of food in a way that is not mediated by the capitalist market (Vergara-Camus, 2014). Thus, from the 1990s, as neoliberalism deepened its adverse social and ecological contradictions, landless rural workers and indigenous peasants have been seeking to gain access to land, or have been defending their right to land, in order to secure the necessary subsistence and a future free from precarity for their families and communities. Under these conditions of deepening precarity and ecological contradiction, a rural to urban exodus is no longer an option for the majority; the demand is primarily for secure access to land in the countryside.

The point to be made here is that, unlike the Global North, a rural to urban exodus has not happened on a permanent basis. Huge numbers of people in the Global South are not full proletarians but semi-proletarians, retaining links to land 'at home'. Moreover, even for the landless, the city no longer represents an 'exit strategy' precisely because of the extreme precarity of employment opportunities across large swathes of the urban South. So, 'right to the city' cannot represent an exclusive strategy but needs to go hand in hand with 'right to the countryside'. It needs to be understood that migration to the city is an expression of the unresolved agrarian question; in terms of taking power, since so many 'proletarians' are actually 'semi-proletarians', there is an almost natural political alliance between city and countryside. Counter-hegemonic forces almost succeeded in Bolivia and Ecuador in the early 2000s, and next time may well succeed in overturning capitalism (see below for details).

The current failure of the city to act as an 'exit strategy' is illustrated well in the words of a member of the Bolivian MST, whose experience could be replicated almost anywhere in Latin America:

> In 1998 I went to El Alto [a huge barrio on the edge of La Paz] to work construction – my brother was there already. When I got there I could not find work, I could not even make enough money to survive. In the city, where I had previously supplemented my family's earnings, now I couldn't even find enough work. So faced with the decision, go back to the countryside, defeated, or stay in the city and try to earn money, I decided: I will go back to the countryside, but not defeated, in fact, ready to struggle, more aware of the sources of our exploitation in the countryside. I will go back and, to ensure a better life, I will find land. … This is the only way to survive, to survive with dignity. (Quoted in Enzinna, 2009, 225)

In Brazil, for example, the sheer size of the rural population expelled from the countryside eliminates the possibility of a meaningful urban 'exit strategy' according to the classic agrarian transition of the Global North (that is, to become urban-based wage workers with relatively secure employment). Again, the sheer size of the informal sector and precarity in urban centres imply that a choice emerges between social marginalisation in the city and land occupation in a countryside where, due to highly concentrated levels of ownership, plenty of land is, in principle, available for settlement by the landless or land poor (Vergara-Camus, 2014). It is clear to such subalterns that expulsion from the land ('primitive accumulation'), migration to the city, and demand for land access in the countryside are part of the same process and problem. Why seek a solution in the city when the problem lies in the countryside? Similarly, in Mexico, migration to the United States has replaced the rural to urban exodus within the country as an exit strategy, migration and exodus induced, of course, by the subsidised 'dumping' of US corn in the Mexican market and the undercutting of *campesino* producers. Ironically then, these migrants seek 'crumbs' from the table of the imperial mode of living in the United States, when that very same mode of living is responsible for forcing them to migrate in the first place. If and when, under the influence of the new Trumpist 'authoritarian populism', that 'safety valve' is closed off to Mexican subalterns, the contradictions of the imperial mode will come home to roost, and we may anticipate a new and more widespread wave of rural and urban unrest, with access to land and the ability to produce food as a use value comprising fundamental demands.

The prospects for food sovereignty as counter-hegemony

Counter-hegemony is taken here to mean opposition to, and autonomy from, the state-capital nexus. 'Radical' food sovereignty, as the ability to produce essential use values unmediated by the capitalist market, comprises a foundational element of such counter-hegemony. Crucial here is how we choose to define capitalism and its relationship to the modern state, because this definition will influence deeply how we envisage counter-hegemony as emancipatory politics.

The definition preferred here derives from the school of so-called Political Marxism, exemplified in the work of Brenner (1977, 1985), Wood (1995, 2009), and Mooers (1991), particularly. This approach has much in common with Gramscian and Poulantzian theory (Poulantzas, 1975), with an emphasis on class dynamics and social-property relationships within the state as key explanatory factors, whilst seeking to situate these dynamics within the wider enabling and constraining political economy of the world capitalist core-periphery structure. Following 'Political Marxism' (see Vergara-Camus, 2014; Tilzey, 2017a, 2018), we can understand capitalism as the contradictory combination of a set of social relationships characterised by:

- the separation, wholly or partially, of workers from their means of production;
- market dependence of producers (the compulsion to depend on a competitive market for the reproduction either of the worker [and family] or of the capitalist enterprise);
- the dominance of absolute private property (the extirpation of common rights in land, and the determination of the right to land only through the capitalist market and, therefore, the alienability of land through its commodification);
- the compulsive imperative of competition among producers (both workers and capitalists);
- the separation of the 'economic' from the 'political' in the form of the modern, capitalist state;
- commodity fetishism;
- the predominance of exchange value over use-value since surplus value extracted from workers is contained in commodities sold on the market.

The loss of the labourer's control over his/her labour power that is entailed in the expropriation of labourers from their means of production through the process of 'primitive accumulation' is, as we have seen, the absolutely key element in the emergence of the above characteristics. It is the obstruction or reversal of this process of primitive accumulation that comprises the absolutely key demand and desire of counter-hegemonic peasant forces in the Global South. This demand and desire persists because the majority of subalterns in the Global South are semi-proletarians, still retaining some access to land, however inadequate this may be (they are *formally*, rather than *really*, subsumed within capitalist relationships of production). Such direct access to non-commodified land exemplifies a form of production in which labourers (semi-proletarians) still control their labour power and, to some extent, the degree and form of integration into the market. Therefore, semi-proletarian peasants, even those who depend on the market for the fulfilment of a significant element of their subsistence needs, have more room for manoeuvre (more 'autonomy'), through the adjustment of production and consumption, than their fully proletarianised counterparts (Vergara-Camus, 2014). Under the prevailing conditions of precarity and 'jobless growth', such autonomy, even if partial, is greatly valued. Unsurprisingly, the aspiration of many under such circumstances is to secure greater autonomy from the capitalist market, in other words, to secure greater access to land in order to achieve self-sufficiency in the production of basic use values such as food. The aspiration, in other words, is to secure a relationship to the market that is one of *opportunity*, not of *compulsion*. Such a condition might be described as one of radical food sovereignty or livelihood sovereignty (Tilzey, 2018).

Given our understanding of capitalism, the achievement of this condition requires the exhaustive transformation and abrogation of capitalist social-property relationships towards democratic and devolved common ownership, or better, stewardship, of the means of livelihood. Again, given our previous discussion, a class-relational or Political Marxian understanding of capitalism appears necessary as a basis for this transformation (see Tilzey, 2017a). Essential prerequisites here are the sundering of the capital-labour relationship and the abrogation of the alienability of property that underlie the competitive impulse of capitalism and the atomisation of non-capitalist collectivities. Since the alienation of land and labour constitute the quintessence of capitalism, it is the reappropriation of land by the dispossessed or partially dispossessed, and the retention of land by those fortunate enough to sustain customary access to it, for the cooperative production of use values for society as a whole, that mark the key elements in capital's transcendence and as the basis for future sustainability. Here the transformation of class-relational power through political action within and around the state (that is, seizing opportunities for autonomy at the local level whilst addressing wider social-property relationships by confronting the state), a 'dual powers' strategy, will be key in expunging exploitative relationships and laying the jurisdictional and material foundations for social equity, cooperative organisation, and ecological sustainability.

If the social relational and institutional transformation of the state-capital nexus is the essential prerequisite for food and livelihood sovereignty, which social forces might realise this? We suggest that it is the middle and lower peasantries, and indigenous peoples, possibly in alliance with the proletarian precariat, which comprise the main counter-hegemonic agent for emancipatory politics as livelihood sovereignty. This is because they view access to non-commodified land, the escape from market dependence, and the equitable and ecologically sustainable production of use values to meet fundamental needs, as the key objectives of social relational transformation. So, although the middle and lower peasantries have indeed become progressively more (semi)proletarianised under neoliberalism, contra Bernstein (2014), they have resisted the adoption of a proletarian class positionality. This is because, for them, poverty equates to a gradual loss of peasant status, which they consequently seek to reverse.

The desire for such a reversal has indeed become ever more insistent as the contradictions of neoliberalism have mounted and the proletariat has increasingly become a precariat. Under these conditions, access to land, however limited, often provides the only real element of livelihood security. Therefore, struggles in the countryside and the city often have an essentially peasant character because disarticulated development cannot provide salaried employment as a viable alternative to secure the means of livelihood. Both peasants and workers seek refuge in the peasant situation, therefore, to maximise the auto-production of use values to meet fundamental needs. The rise of indigenous and ecological consciousness since the 1990s and the de-legitimation of capitalist modernism have served only to reinforce the hunger for land and aversion to full proletarianisation (Moyo and Yeros, 2005). Therefore, the resolution of the unresolved agrarian question of the peasantry in much of the Global South, particularly in the current, and increasingly volatile, ecological conjuncture, seems more than ever to be agrarian and peasant in nature. In this, the potential for mass mobilisation on the part of the middle/lower peasantries, the precariat, and indigenous groups, for an agrarian solution to the contradictions, 'political' and 'ecological', of capitalism does not seem unrealistic. Indeed, the history of "peasant wars" (Wolf, 1999), including the 'second wave' of anti-neoliberal uprisings in Bolivia, Ecuador, Nepal, and elsewhere, indicates ample precedent.

Such mobilisation does not appear unrealistic precisely because capitalism, whether neoliberal or neo-developmental (as in Bolivia and Ecuador, see below), cannot represent a solution to the 'problem' of the peasantry, either through the creation of a class of commercial farmers and/or through the 'real' subsumption of the peasantry as proletarians. Only a fraction of the huge 'peasant' precariat can be absorbed within capitalism as either, for 'political' and 'ecological' reasons. 'Politically', therefore, there has to be a 'peasant' resolution to the agrarian question, since this represents the only realistic social solution to burgeoning precarity. It is precisely because the peasantry is located overwhelmingly in the Global South that a peasant revolution and resolution (to the agrarian question) is likely to occur in the 'periphery'. 'Ecologically', moreover, capitalism cannot represent a resolution because it is both running out of the 'cheaps' that underpin its high productivity/'efficiency' and simultaneously impairing the eco-systemic functions upon which all agricultural production is ultimately founded. The 'peasant way' is, therefore, both a 'political' and an 'ecological' imperative, such that this imperative becomes a political (agro)ecology of praxis, a 'liberation ecology' for the 21st century.

Thus, the 1990s and 2000s saw widespread resistance in Latin America to the socially polarising consequences of neoliberalism and to the progressive loss of national sovereignty (including sovereignty over food) that accompanied neoliberal-dependent 'development' (Veltmeyer and Petras, 2000). Bolivia and Ecuador are exemplary 'pink tide' states where 'radical' forces, comprising middle/lower peasantry, semi-proletarians, proletarians and landless, and indigenous people, engaged in what proved to be a fateful alliance with 'progressive' national bourgeoisie and upper peasantry to displace neoliberalism, only to install national-popular regimes of reformist capitalism. Both states have a new commitment to greater state guidance and interventionism in the economy, to national food sovereignty (albeit largely rhetorical), and to the introduction of social programmes to alleviate the severe income disparities of the neoliberal era. Funds for the latter, however, are predicated on the proceeds of the 'new' extractivism, of minerals, fossil fuels, and agri-fuels, offered by the emergence of sub-imperial states, notably China. These funds have been deployed to subsidise welfarism and infrastructure projects, placating counter-hegemonic constituencies, whose demands for radical land redistribution and land rights remain largely unmet.

In Bolivia, a broad coalition of peasant, indigenous, and worker organisations (*Pacto de Unidad*) succeeded in overthrowing the neoliberal regime in 2003 and installing Evo Morales' MAS (*Movimiento al Socialismo*) party in power from 2005. *Inter alia*, these organisations sought an 'agrarian revolution', entailing massive redistribution of land away from the oligarchy and in favour of the peasantry. However, the success of MAS was also dependent on the support of the national bourgeoisie and the upper peasantry, and it was these constituencies that came to define MAS policies as national-popular capitalism, rather than an (agro)ecologically based socialism. MAS came to be a regime that pursued a sub-hegemonic, or populist, programme of capitalist reformism, placating its counter-hegemonic constituency through welfarism and anti-imperial rhetoric, and soothing the landed oligarchy through accelerated agri-food extractivism and effective exemption from agrarian reform.

The experience of Ecuador is very similar. The period leading up to 2006 saw counter-hegemonic social movements presenting a powerful challenge to neoliberalism. These movements, comprising the *Mesa Agraria*, signed an agreement with the future president Rafael Correa (a middle-class populist), in which he gave a commitment, upon election, to initiate an 'agrarian revolution' based on the peasant/indigenous movement demand for 'radical' food sovereignty, centred on the democratisation of land access, and on state resources for the revival and stimulation of the 'peasant' economy. Again, like Morales in Bolivia, however, it was never Correa's intention to challenge capitalist social-property relations. As in Bolivia, this was to be a national-popular, capitalist-reformist regime. Its populism pivoted on the nationally focussed bourgeoisies' and petty bourgeois class fractions' easy co-optation of the 'progressive' tendency within the food sovereignty movement, and the neutralisation of the 'radical' tendency through social welfare payments and construction projects funded by extractivism. The regime also placated the landed oligarchy by effectively exempting it from the 'agrarian revolution', a revolution, however, that has largely failed to materialise.

These states, in their ability to subvert counter-hegemony through consumerism and welfarism, thus represent, in microcosm (albeit as a pale reflection of), the 'imperial mode of living'. Nonetheless, this ability to subvert and co-opt is time-limited, being dependent on the non-renewable character of neo-extractivism and the ecological despoliation that this entails. With counter-hegemonic demands for an 'agrarian revolution' unmet, the exhaustion of resources and soils through extractivism will presage dwindling funds for the 'compensatory' capitalist state, the unravelling of fragile populist alliances, and a resurgence of 'radical' resistance. This time around, the scarcity of 'ecological surplus' from extractivism will severely curtail the ability of the state-capital nexus to deflect counter-hegemonic forces from seeking an agroecological and peasant-based resolution of the agrarian question – a reversal of 'primitive accumulation' to address both rural and urban precarity through 'livelihood sovereignty' (Tilzey, 2018, 2019).

The experiences of Bolivia and Ecuador serve to demonstrate just how challenging it is for counter-hegemonic social movements to displace the state-capital nexus, even where these movements comprise the majority of the citizenry. This is not to trivialise the important efforts of urban and rural agroecological movements in the Global North, but merely to indicate how much more work is required to 'visibilise' and 'de-reify' the 'imperial mode of living'.

Conclusions: implications of peasant counter-hegemony for urban agroecology

The radical peasant mobilisations described above are important in reclaiming the countryside as the essential agrarian basis for the resolution of the peasant question as the subversion of capitalism, and more particularly of urban capitalism. These mobilisations de-reify the city as the embodiment, or imaginary, of the future as both capitalist and as urban *capitalism*. They problematise not merely the neoliberalisation of the city but, simultaneously, the neoliberalisation of the countryside, the dynamics of the one dialectically related to the dynamics of the other. Similarly, these mobilisations claim not merely the 'right to the city', but, perhaps even more importantly, also the 'right to the countryside', since this is where, overwhelmingly, the resolution to the problem of precarity lies, that is, through reclaiming access to that most fundamental means of production – land. Indeed, these mobilisations presage the strategies identified by Tornaghi (2017) as potentially facilitating the out-scaling of the urban food movement by means of: firstly, stimulating urban agroecology's capacity to challenge neoliberal urbanism; secondly, helping urban and agrarian strategies to converge; and thirdly, experimenting with food commoning.[2] However, they presage these in a more fundamental way by de-reifying the largely urban problematic that is a feature of the Global North, for reasons set out above. If food sovereignty movements can succeed in overthrowing capitalism and the capitalist state in the Global South, the resulting eco-socialism will profoundly subvert the imperialism on which the North's consumer affluence, and its urban-rural dichotomy, are predicated. It will do this because, with eco-socialism, the spigot that feeds surplus value and 'ecological surplus' from the South to the imperial mode of living in the Global North will be turned off. With the resulting material emasculation and discursive deconstruction of urban-based capitalism in the North, pressures from the new, burgeoning (and largely urban) precariat for access to the basic means of production will grow commensurately. These pressures can emerge from current urban agroecological movements, movements that can also, by compromising Northern markets for neo-extractivism, in turn facilitate and strengthen counter-hegemony in the South. Under such circumstances, pressures for direct access to land by this precariat will no longer be confined to the city, however, but, synergising with rural land justice movements, will ramify outwards to the countryside to reclaim the right to land – the land taken from the former peasantries by the process of primitive accumulation. Just as capitalism was born in the countryside (Brenner, 1985), so will its nemesis be found in the reclamation of that countryside by those dispossessed of their agrarian birthright.

Notes

1 These terms have been employed by Wallerstein (1974) and we adopt them here, although our theoretical position differs importantly from his own, that is, World Systems Theory.

2 Deh-Tor (see chapter 1) has since elaborated these strategies for, and definitions of, a post-capitalist political agroecology, and we concur fully with them all. We hope this chapter can complement their argument by locating it within the spatio-historical context of actual state-capitalist dynamics so as to give a broader view of the predisposing circumstances that might engender a shift to this post-capitalist imaginary. By the same token, we seek to afford a strategic relational view on which groups/classes/class fractions are most likely to advocate such a radical programme. Although unpalatable, we take the somewhat pessimistic view that the majority in the Global North will be reluctant to forsake its stake in the imperial mode of living unless forced, by countervailing circumstances, to do so. But we are very happy to be proven wrong!

References

Bernstein, H. (2014) 'Food sovereignty via the 'Peasant Way': a sceptical view', *Journal of Peasant Studies*, vol 41, no 6, pp1031–1063.

Bernstein, S. (2002) *The Compromise of Liberal Environmentalism*. New York: Columbia University Press.

Brand, U. and Wissen, M. (2012) 'Global environmental politics and the imperial mode of living: articulations of state-capital relations in the multiple crisis', *Globalizations*, vol 9, no 4, pp547–560.

Brand, U. and Wissen, M. (2018) *The Limits to Capitalist Nature: Theorizing and Overcoming the Imperial Mode of Living (Transforming Capitalism)*. Lanham: Rowman and Littlefield.

Brenner, R. (1977) 'The origins of capitalist development: a critique of neo-Smithian Marxism', *New Left Review*, vol 104, pp25–93.

Brenner, R. (1985) 'The agrarian roots of European capitalism', in T. H. Aston and C. H. E. Philpin (eds) *The Brenner Debate: Agrarian Class Structure and Economic Development in Pre-Industrial Europe*. Cambridge: Cambridge University Press.

Chibber, V. (2013) *Post-Colonial Theory and the Specter of Capital*. London: Verso.

De Janvry, A., Sadoulet, E. and Wilcox Young, L. (1989) 'Land and labour in Latin American agriculture from the 1950s to the 1980s', *Journal of Peasant Studies*, vol 16, no 3, pp396–424.

Deh-Tor, C. M. (in review) 'Food as an urban question, and the foundations of a reproductive, agroecological, urbanism', *Environment and Planning C (Politics and Space)*.

Enzinna, W. (2009) 'All we want is the Earth: Agrarian reform in Bolivia', in L. Panitch and C. Leys (eds) *Socialist Register 2008: Global Flashpoints, Reactions to Imperialism and Neoliberalism*. London: Merlin Press.

Exner, A., Fleissner, P., Kranzl, L., and Zittel, W. (2013) *Land and Resource Scarcity: Capitalism, Struggle, and Well-being in a World without Fossil Fuels*. London: Routledge.

Harvey, D. (2003) *The New Imperialism*. Oxford: Oxford University Press.

Jessop, B. (2005) 'Critical realism and the strategic relational approach', *New Formations: A Journal of Culture, Theory and Politics*, vol 56, pp40–53.

Kautsky, K. (1988) *The Agrarian Question, 1 and 2*. London: Zwan Publications.

Marx, K. (1972) *Capital, 1 and 3*. London: Lawrence and Wishart.

Mooers, C. (1991) *The Making of Bourgeois Europe: Absolutism, Revolution, and the Rise of Capitalism in England, France and Germany*. London: Verso.

Mooers, C. (2014) *Imperial Subjects: Citizenship in an Age of Crisis and Empire*. London: Bloomsbury.

Moore, J. W. (2015) *Capitalism in the Web of Life: Ecology and the Accumulation of Capital*. London: Verso.

Moyo, S. and Yeros, P. (eds) (2005) *Reclaiming the Land: The Resurgence of Rural Movements in Africa, Asia and Latin America*. London: Zed Press.

Moyo, S. and Yeros, P. (eds) (2011) *Reclaiming the Nation: The Return of the National Question in Africa, Asia, and Latin America*. London: Pluto Press.

Perelman, M. (2000) *The Invention of Capitalism: Classical Political Economy and the Secret History of Primitive Accumulation*. Durham: Duke University Press.

Poulantzas, N. (1975) *Classes in Contemporary Capitalism*. London: New Left Books.

Smith, J. (2016) *Imperialism in the Twenty-First Century: Globalization, Super-Exploitation, and Capital's Final Crisis*. New York: Monthly Review Press.

Tilzey, M. (2016) 'Global politics, capitalism, socio-ecological crisis, and resistance: Exploring the challenges and the linkages', Colloquium Paper No 14, Global Governance/Politics, Climate Justice and Agrarian/Social Justice: Linkages and Challenges: An International Colloquium, 4–5 February 2016. ISS, The Hague.

Tilzey, M. (2017a) 'Reintegrating economy, society, and environment for cooperative futures: Polanyi, Marx, and food sovereignty', *Journal of Rural Studies*, vol 53, pp317–334.

Tilzey, M. (2017b) 'Market civilization' and global agri-food: understanding their dynamics and (in)coherence through multiple resistances', in A. Bonanno and S. Wolf (eds) *Resistance to the Neoliberal Agri-Food Regime: A Critical Analysis*. New York: Routledge.

Tilzey, M. (2018) *Political Ecology, Food Regimes, and Food Sovereignty: Crisis, Resistance, and Resilience*. London: Palgrave Macmillan.

Tilzey, M. (2019) 'Authoritarian populism and neo-extractivism in Bolivia and Ecuador: The unresolved Agrarian question and the prospects for food sovereignty as counter-hegemony', *Journal of Peasant Studies*, vol 46, no 3, pp626–652. doi: 10.1080/03066150.2019.1584191.

Tornaghi, C. (2017) 'Urban agriculture in the food-disabling city: (re)defining urban food justice, reimagining a politics of empowerment', *Antipode*, vol 49, no 3, pp781–801.

Veltmeyer, H. and Petras, J. (2000) *The Dynamics of Social Change in Latin America*. London: Palgrave Macmillan.

Vergara-Camus, L. (2014) *Land and Freedom: The MST, the Zapatistas and Peasant Alternatives to Neoliberalism*. London: Zed Press.

Wallerstein, I. (1974) *The Modern World System: Capitalist Agriculture and the Origins of the European World Economy in the Sixteenth Century*. Vol. 1. New York: Academic Press.

Wolf, E. R. (1999) *Peasant Wars of the Twentieth Century*. Norman: University of Oklahoma Press.

Wood, E. M. (1995) *Democracy against Capitalism: Renewing Historical Materialism*. Cambridge: Cambridge University Press.

Wood, E. M. (2002) 'The question of market dependence', *Journal of Agrarian Change*, vol 2, no 1, pp50–87.

Wood, E. M. (2009) 'Peasants and the market imperative: the origins of capitalism', in A. H. Akram-Lodhi and C. Kay (eds) *Peasants and Globalization: Political Economy, Rural Transformation, and the Agrarian Question*. Abingdon: Routledge.

CONCLUSIONS

The programmatic dimension of an agroecological urbanism

Michiel Dehaene and Chiara Tornaghi

Introduction

After having taken the reader on a journey along different methodologies, concepts and political stances, in these conclusions we want to come back to our original aim of foregrounding an agroecological urbanism as a paradigmatic change for thinking sustainable urban food environments and collective solidarities. To this goal, we proceed here with a discussion of what we see as the key dimensions of its programmatic nature.

In Sections 1–4, we begin with reinstating that an agroecological urbanism (i) speaks across the urban-rural divide, (ii) defines a post-capitalist imaginary for urbanism, (iii) depends on politicised pedagogies to move it forward and (iv) is action-oriented and seeks to engage in concrete places.

In Section 5, we then illustrate some of the progress we have made in our current research work, searching for concrete tools to build an agroecological urbanism with different communities of practice. A fuller illustration of this work will be at the centre of our next book.

We then conclude in Section 6 with a reflection on how the COVID-19 pandemic that is unfolding while we are writing these conclusions has strengthened and rendered even more urgent the transformative ambition of an agroecological urbanism.

1. Speaking across the urban-rural divide

While the notion of an agroecological urbanism that we have illustrated (cf. Chapter 1 in this book, as well as Deh-Tor (2017)) is intended as an expansive and inclusive concept, in this book it has, first and foremost, been projected as a conversation between two emerging communities: the political agroecology community on the one hand, and the sustainable food planning community on the other. Political agroecology moves from a strong and coherent agenda rooted in agrarian struggles and in the value positions of agroecology; however, it looks at the city and dynamics of urbanisation as a direct threat (Tornaghi and Dehaene, 2020). Sustainable food planning has convincingly demonstrated how the focus on food can act as a catalyst to think about the sustainable transformation of cities, but to date it largely remains an heterogeneous and weakly politicised movement mostly confined within a selective understanding of the (urban) food question centred on food consumption. It has no tradition nor any specific language to speak to farmers. Given the long history of separation, not only between specific geographical settings, but also between different worlds, with different constituencies, different interests and different subjectivities, such a conversation between agroecology and food planning is, in fact, hard to hold. The proposition of an agroecological urbanism is not just a matter of putting these worlds together. It is an adventure that seeks the transformation of both. With this book we have begun, from an agroecological point of view, a systematic interrogation of the way cities have been organised. Agroecology gives food planning a normative point of reference to define the urban transformations to which it aspires. However, we also ask agroecology to imagine its future on a highly urbanised planet, within an urbanised society. With the latter we do not mean 'within the city', but rather within a context of irreducible social differences, within a context of exacerbated interdependence. An agroecological urbanism then seeks to imagine forms of place-based solidarity and collective arrangements within the diverse and interconnected demographics of the planetary urban.

The effort to speak across the urban-rural divide is first and foremost an attempt to question a history of capitalist urbanisation that has normalised the territorial separation of the urban and the agrarian question. While this history has precedents in the emergence of the mercantile city, it really took its systemic and systematic form under dynamics of industrialisation. This led to an urbanism that installed the needed food processing and transport logistics that would structure the rural-urban interface in terms of the exchange of food in commodified forms (Cronon, 1992). It also enabled the logics that made it possible, for the time being, to squander fertile soils for urban expansion.

Against this historical background, more is needed than the reconnection of urban consumers and rural producers, more also than the promotion of urban agriculture that has the merit of reintroducing forms of food production within urban contexts from which food production has been systematically removed. Urban agriculture insufficiently questioned the intimate relationship between urbanisation and the excessive commodification of food in the urban context. In the worst case, the promotion of soil-less forms of food production under the urban agriculture label, while presented as a way of saving land, in fact provides the perfect excuse for the destruction of resources (soil to begin with) that are fundamental to agroecological food production.

The ethical principles of agroecology, rooted in environmental justice, inter-species solidarity, principles of environmental care and stewardship provide all the cues to reconnect urban lives to models of food production that regenerate the ecological basis on which these urban lives depend. These principles, however, have not historically been directed at the construction of subjectivities and social movements that embrace the urban. Taking forward, an agroecological urbanism will require solidarities and, ultimately, the construction of new collective subjectivities between urban and agrarian movements (Tornaghi and Halder, forthcoming).

2. Defining a post-capitalist imaginary

Perhaps even harder than imagining a world that is not sorted along urban and rural lines is to envision urbanisms that are not the formal translations of the political economy of capitalist urbanisation. To a large extent, not only has the critical interpretation of the history of urbanisation been designed to debunk the recent history of urbanisation as the expression of capital accumulation and as the spatial fix of its internal contradictions (Harvey, 1985), the long history of urbanisation in the West runs historically parallel with the history of capital. It is, however, to that same tradition that we owe the call for an alternative urbanism: 'a genuinely humanising urbanism ... yet to be brought into being' (Harvey, 1973). The effort to imagine an agroecological urbanism is an attempt to answer that call and to render concrete what a post-capitalist urbanism could look like.

Arguing that urbanism and capitalism are not two sides of the same coin may seem superfluous and removed from all empirical evidence. And yet, through an agroecological urbanism, we wish to join the ranks of those that try to think of a future beyond capitalism. An agroecological urbanism seeks to break with the extractive and colonial logics upon which the history of urbanisation has been built, and to incorporate the ethics of the regeneration of resources within the urban. The focus on food and food production implies a humble position of planning and needs to go hand in hand with the acknowledgement of its historical complicity with a structural neglect of the urban food question, and with the progressive marginalisation of local food provisioning practices that historically maintained a connection between the city and its *terroir*.

The pervasive presence of food in our capitalist daily lives makes it a rich subject to engage in the methodologies of Gibson-Graham, namely to (re)learn to see the diversity that is out there and is not accounted for by the logics of capitalism (Gibson-Graham, 2006). An agroecological urbanism may build on all the food provisioning practices that have been residualised by capitalist urbanisation. Documenting the many practices that still exist, and are still reproduced by local communities or indigenous cultures, not only breaks the totalising spell of capitalism, but also provides direct evidence that other ways of organising exist and may be outscaled and empowered. Moreover, such attention to diverse economies and alternative sets of values is equally critical to devise solidarity strategies to live with the irreparable damage of the legacy of capitalism.

Taking stock of the ways in which communities have been forced to live within difficult and resource-deprived situations, we may find concrete cues for thinking more systematically about what it entails to live on a damaged planet (Tsing, 2017). We are thinking, for example, of the efforts of cities in the global south to reclaim damaged soils for food production through the rebuilding of topsoils with food and green waste.

3. The role of politicised pedagogies

Politicised pedagogies are essentials to the project of an agroecological urbanism in all its transformative ambitions. As the agroecological movement and scholarship has highlighted, the development of agroecology has been consistent with the strengthening of communities of practices, the building of new subjectivities and the articulation of political movements. The marriage of agroecology and food sovereignty movements has marked an even more consistent pathway of politicisation for agroecology and more direct attempts to address food system transformation. Practice and scholarship on agroecological pedagogies, however, has focussed mostly on farmer-to-farmer (peasant-to-peasant) learning and farmer training schools inspired by Freirian pedagogical models; they have also focussed on rural contexts and actors (McCune and Sánchez, 2019). This focus on the rural and the active disengagement with the urban realm has been detrimental to the capacity to see the urban as a frontier of struggle, a source of alliances and an arena to be reclaimed by the movement (Tornaghi and Dehaene, 2020). The focus also acknowledges that transforming the food system requires politics apt for transforming processes of urbanisation and urbanised ways of life at the same time.

The way ahead, and the programmatic point for an agroecological urbanism, is therefore to rethink the political pedagogies of the agroecological movement in a way that is fit to address the challenge posed by current processes of urbanisation and the residualisation of agroecological farming in the urban and peri-urban context. It requires questioning whether existing pedagogies in urban contexts are sufficiently specific to equip the farmers with the knowledge to navigate urban challenges of setting up viable businesses (i.e., to address the specific land access, nutrient sourcing, economic models, intersectional solidarities and consumer engagement needed to thrive within a specific geography or marginalisation).

It requires fostering an ecology of learning and an ethic of care, able to trigger deep value-shifts across different urban communities and to build intersectional solidarities. It also requires turning solidarities across constituencies in the locality into new political subjectivities that would be able to articulate a transformative agenda and prepared to organize and lobby for its implementation.

4. An action-oriented and territorially grounded practice

Agroecology as a practice, movement and science (Wezel et al., 2009) is embedded in concrete communities and places, and it strongly values indigenous and local knowledge. It comes with an acknowledgement of contextual geographical, social, historical and cultural difference. An agroecological urbanism appeals to this ethos to transform the field of urbanism, to decolonise a discipline permeated with the dialectics of enlightenment and rational domination, in favour of a distributed and transdisciplinary knowledge ecology. It believes in the role of farmers and food growers as stewards of local resources, as key players in the reproduction of knowledge and skills, and leading agents of a food system change. It looks at the urbanised landscapes as 'peopled landscapes' (Ward, 1999), inhabited by bodily subjects who owe their livelihoods to the context of which they are a part.

An agroecological urbanism is a concrete and action-oriented agenda. It is about local action in particular places, where a concrete agenda for urban agroecology could be shaped as part of a radical commitment to the urban, i.e., a radical commitment to shaping place-based solidarities and to a social contract shaped around the acknowledgement of the interdependence of people living their lives within the same territorial basin. These can be the community-driven relationships of a neighbourhood or urban district, but they can also include the wider solidarities and collective infrastructures to be built at the level of the agglomeration. An agroecological urbanism challenges planners to expand their horizon beyond the limits of an urban project that thinks such arrangements mostly around housing, transport and energy infrastructure, and begin to see community kitchens, a shared composting infrastructure, green-blue networks, water harvesting infrastructure, etc. as essential parts of the way that an ecology-caring urbanisation takes form.

5. Eight steps forward: advocacy planning for an agroecological urbanism

An agroecological urbanism model is not ready to be implemented; rather it is an agenda for the systematic reskilling and retraining of communities to transform the food-disabling city into a food-enabling one. We imagine that the needed learning in which people will have to engage could take the form of advocacy planning through which concrete communities of practice begin to define ways to systematically embed agroecological food growing into an urban environment. We imagine, for example, how the emerging community working around questions of soil care (cf. Soil Care Network) would engage in the translation of concrete engagement with the soil into a new narrative for spatial planning. We imagine how existing groups engaged in the use of urban waste in composting could lead to the identification of specific nutrient sources within an urban environment and ways in which they could be collected, processed and applied in the remediation of damaged urban lands. We imagine how community kitchen initiatives would team up and systematically build the needed environments to self-produce local food and share it within the communities that support and host them. We imagine how local farmers already mobilised around the issue of land access would question public land policies and imagine a new generation of land readjustment programmes that valorise the residualised soils of the peri-urban fringe and rebuild relationships of nutrient exchange between farmland, pasture land and the wooded parts of the urbanised landscape.

As part of the *Urbanising in Place* project, we have been working with communities in Brussels, London, Rosario and Riga, identifying concrete opportunities for the development of an agroecological urbanism (Tornaghi and Dehaene, 2020). In an effort to bring the lessons learnt to an international forum, the consortium behind this programme started to identify eight building blocks of a transformative agenda to refashion the way we organise cities on an agroecological basis. These blocks identify concrete points of articulation that, today, are missing from the urban landscape, or exist only in marginalised forms. The blocks are intended as descriptors of concrete matters of concern, transformational practices and political landscapes around which local and international conversations between agroecologists and sustainable food planners are being set up. We are working to give an actual account of these conversations and the lessons learnt in a second book. The eight building blocks focus on different and complementary territorial scales, but all seek to define a specific relational setting and fruitful combinations between the capabilities of concrete actors and the urban conditions in which they are operating. We see these blocks as steps ahead in the imagination of a form of advocacy planning to be developed together with the agroecological community, systematising and conceptualising the contours of an agroecological urbanism. We discuss them in turn below.

The **productive housing estate** looks at forms of urban development that incorporate food productive spaces within housing schemes and thereby make a direct connection between the right to grow (Tornaghi, 2017) and the right to shelter. This block makes a direct connection between food and housing, two dimensions central to the ways in which social reproduction is structured in an urban environment.

The **territorial food hub** mobilises the place-based solidarities that may exist at the district or neighbourhood scale, combining food production with educational activities, social entrepreneurship and community work, in light of multiplying people's resourcefulness and the collective care of resources in a food-democracy perspective.

The **agroecological park** makes a dent in the consumption of fertile soils by urban expansion and reserves a territorially demarcated area for agroecological food production at the interlink between urban and peri-urban fringes. This building block projects dedicated areas for agroecological food production equipped with the collective infrastructure (shared composting facilities, shared marketing infrastructure, technical assistance, common pasture land, etc.) on which individual growers could rely.

The building block **farming the fragmented land** looks at practices that valorise residual patches of agricultural land within the complex land mosaic of the peri-urban fringe. It looks at specific business models, strategies to combine land, specific cultivation choices, etc. that build on the potential use value of fragmented landscapes, and in particular their role in building multifunctional and ecologically interdependent agroecological farming initiatives, which current market and land access conditions have rendered monofunctional.

Through the land-based **community kitchen**, we try to imagine a city in which neighbourhoods would be equipped with a food infrastructure in the same way as they now have elementary schools or health centres. We imagine land-based neighbourhood kitchens that host community composting schemes and food re-skilling sessions, that would be linked-up with growing spaces in the fringe and provide the social infrastructure for a neighbourhood-driven transformation and rescaling of the urban food system.

Politicised pedagogies, as discussed above, are an essential component of an agroecological urbanism. This building block structures the positioning work, the alliance building and the personal transformation necessary to install a new relational geography within the hearts and minds of new agroecology-minded urbanites.

The **land and market access incubator** combines the training of a new generation of professional farmers with the provision of testing spaces and the facilitation of land access within a competitive urban land market.

The **healthy soil scape** relates the practices of soil care to a landscape geography in which nutrient streams can be circulated and combined within a balanced ecology of permanent grassland, woodland and arable land. It provides the necessary collective infrastructure and knowledge support to regenerate soil fertility beyond the boundaries of the farm.

These blocks are not intended as an end point or as an attempt to box the project of an agroecological urbanism in a set of formulas. They are intended as a starting point for further experimentation and collective learning. They define a provisional attempt to describe a set of points of 'articulation' at the intersection of agroecology and urbanism. They require political work to turn them into the tools of a transformative process.

6. COVID-19 as a catalyst

The writing of the conclusions of this book took place under COVID-19 lockdown. The challenges of working and caring for ourselves and others during the unfolding of the pandemic and the constraints of lockdown work made us despair as far as the disruption of our writing schedules was concerned. It made us hopeful, however, with respect to the agenda of this book, which only seems to have gained in relevance. Three points in particular make us believe that the pandemic not only makes the current global food crisis tangible, but also creates more room to get a discussion on urban agroecology off the ground.

First, COVID-19 offers us an X-ray of how questions of care and social reproduction are structured in societies. They show what there is and what there is not to fall back on. It presents us with the downside of all the things that have been outsourced within the neo-colonial globalised world order. It places the groups that have traditionally benefited from such asymmetrical, extractive relationships at risk, showing the vulnerability of the global cheap food regime. However, it also creates new opportunities to rebuild diverse economies, to reinvest in residualised arrangements of care, to rebuild skills locally and to shape novel collective solidarities around the transformation of food into meals.

Second, COVID-19 gives new sharpness to the debate on reterritorialisation. While the main driver of this discussion in food planning has been centred around climate change goals and the need to reduce food miles and related carbon emissions, COVID-19 literally disrupted the global supply chains. COVID-19 makes living within boundaries concrete. The discussion on local food production is suddenly not just a tool for something else, an instrument to deliver ecosystem services, but a legitimate subject in its own right. Resource sovereignty under COVID-19 is not an abstract concept, or the sole aim of rural movements, but a direct concern of millions of people who never thought about the link between food and resources before. Within the local food systems, the difference between the high-tech rooftop greenhouse that went out of business as it was only delivering herbs and microgreens to the local catering industry, and the agroecological farm of the CSA (Community Supported Agriculture) that has been honouring principles of farmers' autonomy all along and proved to be extremely resilient in times of crisis became striking.

Lastly, COVID-19 made clear how much the biodiversity crisis is an exponent of the deep connection between agriculture and urbanisation. The pandemic of the past year originated in ecologies of disruption and extraction that bring humans in an invasive manner into wildlife habitats that they render unstable while creating unprecedented and unpredictable exchanges between their respective ecologies (Wallace et al., 2020). These aggressive forms of extraction are not only there to serve the land needs of rapidly expanding processes of urbanisation: they are also connected to resource and migration flows towards these urban habitats. While attention first goes to investing in ways of containing the disastrous effects of ecological disruption, it is time to deeply question the way in which we continue to think urban-nature relationships in utilitarian terms. Urban political agroecology presents an interesting trajectory to begin to reimagine and redefine agroecology-centred ways of life in a highly urbanised world.

Acknowledgements

Section 5 of this chapter is partially built on the preliminary results of an ongoing research project funded under the JPI Urban-Europe-Belmont

Forum-SUGI-NEXUS Programme titled 'Urbanising in Place' (Project number: 11326801). This includes research funding from the following institutions and related grant numbers: FWO (G0H5817N), ESRC (ES/S002251/1), Innovate UK & ERA-NET (620145 & 11326801), NWO (438-17-406), VIAA (ES RTD/2018/15 and ES RTD/2018/16), Innoviris (RBC/2018-ENSUGI-1) and MINCYT(CONVE-2019-16850590-APN-DDYGD#MECCYT).

References

Cronon, W. (1992) *Nature's Metropolis: Chicago and the Great West*. New York: Norton.

Deh-Tor, C. M. (2017) From Agriculture in the City to an Agroecological Urbanism: The Transformative Pathway of Urban (Political) Agroecology. *Urban Agriculture Magazine*, special issue on "Urban Agroecology", no 33, pp.8–10.

Gibson-Graham, J. K. (2006) *A Postcapitalist Politics*. Minneapolis: University of Minnesota Press.

Harvey, D. (1973) *Social Justice and the City*. Baltimore: Johns Hopkins.

Harvey, D. (1985) *The Urbanization of Capital*. Baltimore: Johns Hopkins.

McCune, N. and Sánchez, M. (2019) Teaching the territory: agroecological pedagogy and popular movements. *Agriculture and Human Values*, vol 36, no 3, pp.595–610.

Tornaghi, C. (2017) Urban agriculture in the food-disabling city: (re) defining urban food justice, reimagining a politics of empowerment. *Antipode*, vol 49, no 3, pp.781–801.

Tornaghi, C. and Dehaene, M. (2020) The prefigurative power of urban political agroecology: rethinking the urbanisms of agroecological transitions for food system transformation. *Journal of Agroecology and Sustainable Food Systems*, vol 44, no 5, pp.594–610.

Tornaghi, C. and Halder, S. (2021, forthcoming) Reconfiguring the intersection between urban food movements and agrarian struggles: building an urban political agroecology praxis. In: Engels, B., Dietz, K., Akram-Lodhi, H. and McKay, B. (eds), *The Edward Elgar Handbook of Critical Agrarian Studies (CAS)*, Edward Elgar Publishing.

Tsing, A. (2017) *Arts of Living on a Damaged Planet*. Minneapolis: University of Minnesota Press.

Urbanising in Place project: www.urbanisinginplace.org.

Wallace, R., Liebman, A., Chaves, L. F. and Wallace, R. (2020) COVID-19 and circuits of capital. *Monthly Review*, https://monthlyreview.org/2020/05/01/covid-19-and-circuits-of-capital/.

Ward, C. (1999) A peopled landscape. In: Warpole, K. (ed.), *Richer Futures: Fashioning a New Politics*. London: Earthscan, pp.83–98.

Wezel, A., Bellon, S., Doré, T., Francis, C., Vallod, D. and David, C. (2009) Agroecology as a science, a movement and a practice. A review. *Agronomy for Sustainable Development*, vol 29, no 4, pp.503–515.

INDEX

Page numbers in **bold** denote tables, in *italic* denote figures

For Product Safety Concerns and Information please contact our EU representative GPSR@taylorandfrancis.com Taylor & Francis Verlag GmbH, Kaufingerstraße 24, 80331 München, Germany

Batch number: 10725798

Printed by Printforce, the Netherlands